Approximate Methods in Structural Seismic Design

RELATED BOOKS FROM E & FN SPON

Behaviour of Steel Structures in Seismic Areas
Edited by F.M. Mazzolani and V. Gioncu

Concrete under Severe Conditions: Environment and Loading
Edited by K. Sakai, N. Banthia and O.E. Gjørv

Earthquake Engineering
Y.X. Liu, S.C. Liu and W.M. Dong

Earthquake Resistant Concrete Structures
G.G. Penelis and A.J. Kappos

**International Handbook of Earthquake Engineering Codes,
Programs and Examples**
Edited by M.H. Paz

Natural Risk and Civil Protection
Edited by T. Horlick-Jones, R. Casale and A. Amendola

Nonlinear Dynamic Analysis of Structures
Edited by K.S. Virdi

**Nonlinear Seismic Analysis and Design of Reinforced
Concrete Buildings**
Edited by P. Fajfar and H. Krawinkler

Reliability and Optimization of Structural Systems
Edited by R. Rackwitz, G. Augusti and A. Borri

Soil Liquefaction
M. Jefferies and K. Been

Structural Design for Hazardous Loads
Edited by J.L. Clarke, F.K. Garas and G.S.T. Armer

Structures Subjected to Dynamic Loading: Stability and Strength
Edited by R. Narayanan and T.M. Roberts

Theory and Design of Seismic Resistant Steel Frames
F.M. Mazzolani and V. Piluso

*For more information, contact the Promotion Department, E & FN Spon, 2–6
Boundary Row, London SE1 8HN, Tel: Intl + 44 171-865 0066, Fax: Intl
+ 44 171 522 9623.*

Approximate Methods in Structural Seismic Design

ADRIAN S. SCARLAT
Manager
Research and Technology
A. Scarlat, N. Satchi Engineers Ltd
Tel Aviv, Israel

and

Affiliate Professor
Technion
Haifa, Israel

CRC Press
Taylor & Francis Group
Boca Raton London New York

CRC Press is an imprint of the
Taylor & Francis Group, an **informa** business

A CHAPMAN & HALL BOOK

CRC Press
Taylor & Francis Group
6000 Broken Sound Parkway NW, Suite 300
Boca Raton, FL 33487-2742

First issued in paperback 2019

© 1996 Adrian S. Scarlat
CRC Press is an imprint of Taylor & Francis Group, an Informa business
Typeset in 10/12 Times by Thompson Press (India) Ltd, Madras

No claim to original U.S. Government works

ISBN-13: 978-0-419-18750-9 (hbk)
ISBN-13: 978-0-367-44875-2 (pbk)

A catalogue record for this book is available from the British Library

Library of Congress Catalog Card Number: 95-74654

Visit the Taylor & Francis Web site at
http://www.taylorandfrancis.com

and the CRC Press Web site at
http://www.crcpress.com

Dedicated to
My wife Evemina
My sons Alex, Dan and Yuval

Contents

Contents

Preface

Three distinct stages may be defined in the structural design of buildings:

1. selection of the structural solution and assessment of the dimensions of the main elements;
2. detailed computation of stresses and displacements;
3. checking of the main results.

Note that computerized analysis suits only stage 2, whereas approximate methods have to be used for stages 1 and 3; this points out the exceptional importance of approximate methods in structural design.

This book deals with the most difficult problem arising in the design of multi-storey buildings and similar structures: the effect of horizontal forces, mainly seismic loads and similar loads such as wind pressure. In fact, the general stability of a structure capable of resisting horizontal forces is also ensured for vertical loads; possible accidents are generally only local. Unfortunately, the most difficult task is to define clearly the concept of approximate methods. Their reverse, the accurate methods, also elude definition: in order to apply them, several simplifying assumptions are required. Hence we are obliged to accept a rather vague definition: an approximate method in structural analysis is a method that permits the assessment of stresses or displacements in a much shorter time than the commonly used design methods.

We shall refer in the following to two different types of approximate methods: (1) methods designed to determine the stresses and displacements of a given structure by using substitute structures; (2) methods based on 'global parameters' (seismic coefficients, total area of structural walls, etc.). The advent of computerized structural analysis and, in particular, of finite element programs, has opened up vast possibilities for a fundamental reappraisal of existing approximate methods, including a more accurate definition of their scope and limits. We have taken advantage of these possibilities, have examined several 'classical' approximate methods, and have proposed a number of new techniques intended to complement the existing ones.

This book is the result of many years spent in structural design and teaching. One of its main sources is a course in structural dynamics taught by Professor P. Mazilu at the Institute of Civil Engineering of Bucharest, Romania, which the author was privileged to attend.

The target audience of the book is first of all design engineers, but it should be of use for non-specialist engineers too; it is assumed that it will serve also as a teaching aid for undergraduate students as well as for advanced high-rise buildings courses. It is hoped, by citing Mozart (*toute proportion gardée*), that 'here and there are sections that only connoisseurs will enjoy, but these sections have been written so that even a layman will have to enjoy them, albeit without knowing it.'

Adrian S. Scarlat

ACKNOWLEDGEMENT

The author is especially grateful to Professor A. Rutenberg for his careful screening of the manuscript and for suggesting valuable improvements.

1 Multi-storey building frames

1.1 Introduction

In this chapter we propose to assess the bending moments and the deflections of a multi-storey building frame acted upon by lateral forces (Figure 1.1a), by using approximate methods.

As will be shown in the next section (1.2), the approximate analysis of this frame may be performed on a 'substitute' (equivalent) one-bay, symmetrical frame (Figure 1.1b). Therefore we shall deal mainly with this latter frame, as a first step in analysing the actual multi-storey, multi-bay frame.

In our analysis, we assume several hypotheses aimed at simplifying the computation. These are as follows.

- It is assumed that all the horizontal loads are concentrated at floor levels.
- The effect of the shear forces (V) on the deformations is neglected. This hypothesis is acceptable as long as we deal with the usual systems of bars. The analysis of structural walls, where this effect is important, will be dealt with separately (Chapter 2).
- The effect of axial forces (N) on the deformations is neglected. This hypothesis, too is acceptable as long as the total length of the multi-bay frame (L) is not small with respect to its total height (H). In most practical cases this assumption is justified. We point out, in order to be consistent, that we also have to neglect the effect of axial forces in the analysis of the substitute frame. This means that we have to assume cross-sectional areas

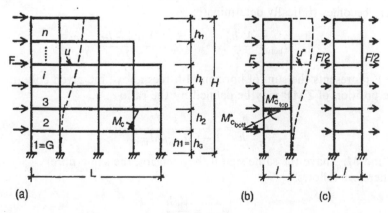

Figure 1.1

$A \to \infty$ (independently of the ratio L/H). Consequently, the asymmetrical load of Figure 1.1b may be replaced by the anti-symmetrical load shown in Figure 1.1c.

In the following, we shall deal mainly with columns fixed in the foundations. The case of pin-supported frames will be considered separately (section 1.2.4). The effect of soil deformability is dealt with separately, too (section 1.3.3).

1.2 Multi-storey, one-bay frames

1.2.1 GENERAL APPROACH

The approximate analysis of multi-storey, one-bay, symmetrical frames subjected to horizontal loads is performed by one of the following procedures:

1. the zero moment point procedure;
2. the continuum procedure (replacement of the beams by a continuous medium). This latter technique is only suitable for multi-storey frames with a large degree of uniformity.

In the case of very irregular substitute frames (from the aspect of both geometry and rigidity), it is advisable to perform the analysis by computer (the 'approximation' will stem from the conversion of the substitute frame to the actual one). We have to assume in this case, too, that the effect of axial forces is negligible (by assuming $A \to \infty$).

1.2.2 THE ZERO MOMENT POINT (ZMP) PROCEDURE

Let us consider the one-bay symmetrical frame shown in Figure 1.2. It is assumed that we know the position of the point where the moment diagram M intersects the column's axis (**zero moment point**, ZMP), i.e. the height h_0. The problem becomes statically determinate:

$$M_{\text{bott}} = \frac{V h_0}{2}; \qquad M_{\text{top}} = \frac{V(h_0 - h)}{2} \tag{1.1}$$

where V represents the sum of horizontal forces above the given column.

The position of ZMP may be defined by the ratio

$$\varepsilon = \frac{h_0}{h} = \frac{1}{1 - (M_{\text{top}}/M_{\text{bott}})} \tag{1.2a}$$

(M_{bott} and M_{top} have the same sign if they tension the same fibre).

It therefore follows:

$$M_{\text{bott}} = \frac{V\varepsilon h}{2}$$

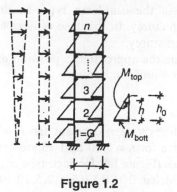

Figure 1.2

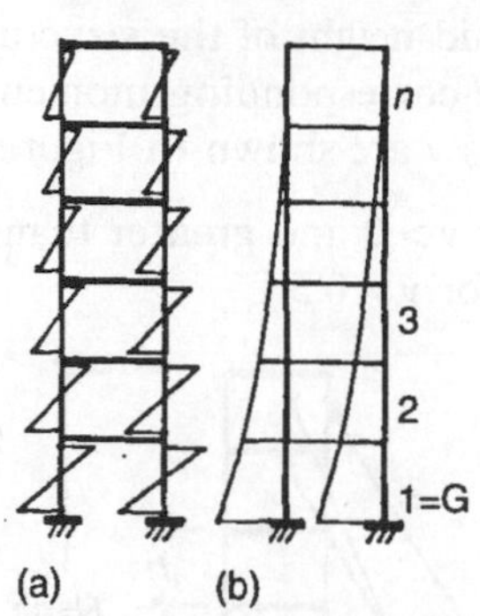

(a)   (b)

Figure 1.3

$$M_{\text{top}} = \frac{-V(1-\varepsilon)h}{2} \qquad (1.2b)$$

For a frame of given geometry and for a given storey i, ε depends mainly on the ratio

$$v = \frac{k_{b_i}}{k_{c_i}} \qquad (1.3)$$

where $k_{b_i} = I_{b_i}/l$; $k_{c_i} = I_{c_i}/h$; and $I_b(I_c)$ are moments of inertia of beam i (column i). We note that in most practical cases the ratio v lies between 0.1 and 5.

The type of loading has some effect on the position of ZMP, but we may safely neglect this effect and assume that for any laterally distributed load ZMP is invariant.

Figure 1.3 presents diagrams for two extreme situations. Figure 1.3a shows extremely stiff beams ($v \to \infty$), where ZMP lies at the mid-height of each storey ($\varepsilon = 0.5$). Figure 1.3b illustrates extremely flexible beams ($v \to 0$); the diagram of bending moments M is of the **cantilever type**. In the case of stiff beams ZMP lies within the given storey. In the case of flexible beams ZMP may be positioned above the given storey.

We shall now determine the approximate position of ZMP by referring to two cases: uniform and non-uniform frames.

(a) Uniform frames

Uniform frames have equal heights, I_c = constant and I_b = constant at each storey.

Analyses were performed for $n = 6, 10$ and 15 storeys subjected to uniform and inverted triangular loads (Figure 1.4); for each type of frame and loading, eight ratios $v = k_b/k_c$ were considered: $0.01, 0.1, 0.5, 1, 2, 5, 10, 1000$ (a total of 48 cases).

Ratios ε were computed at three levels: $\varepsilon_1 = \varepsilon_G$ (at the ground floor), ε_2 (one floor above) and ε_m (at the mid-height of the structure). The ratio ε at the top floor is not significant, as the corresponding moments are usually very small.

Average curves for ε versus v are shown in Figure 1.5 for $\varepsilon_1 = \varepsilon_G$, ε_2 and ε_m:

- ε_G is close to 0.5–0.6 for $v > 2$ and greater than 1 for $v < 0.2$.
- ε_m remains close to 0.5 for $v > 0.5$.

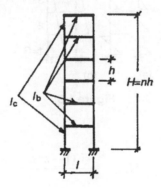

Figure 1.4

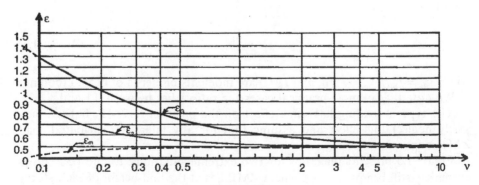

Figure 1.5

We note that even for reinforced concrete structures the column reinforcement is usually uniform along the height of the storey. In cases where $\varepsilon_m \neq 0.5$ (either $\varepsilon_m > 0.5$ or $\varepsilon_m < 0.5$), the maximum moment is greater than the moment we have computed on the assumption that $\varepsilon_m = 0.5$.

Therefore, in design it is advisable to consider the bending moments around the mid-height of the frame, increased by 10–20% with respect to the moments based on $\varepsilon_m = 0.5$.

ε_2 lies between ε_G and ε_m; for $v = 0.1$–0.3, ε_2 varies between 0·5 and 0·9.

For ratios $v < 0.1$, the spread of results is too wide to permit a reasonable average value to be accepted. In such cases, it is more convenient to relate the maximum moment acting on the columns (M_{max} above the foundations) directly to the maximum moment acting on a cantilever (M_{cant}), due to loads $F/2$; see Figure 1.6.

From the frames we have analysed we obtain:

$$
\left.
\begin{aligned}
v = k_b/k_c = 0{\cdot}001 \quad & M_{max}/M_{cant} = \quad 0{\cdot}8\text{--}0{\cdot}9 \\
0{\cdot}01 \quad & 0{\cdot}4\text{--}0{\cdot}6 \\
0{\cdot}1 \quad & 0{\cdot}15\text{--}0{\cdot}3
\end{aligned}
\right\} \tag{1.4}
$$

We note that even for very flexible beams ($v = 0{\cdot}01$) the effect of the beams remains significant (it ensures a decrease of the maximum moments acting on the columns by $\sim 50\%$ with respect to the cantilever moments).

(b) Non-uniform frames

The position of ZMP for several cases of 10-storey frames has been examined. The results are as follows.

- The moments of inertia of the columns vary along the height of the frame (Figure 1.7), the beams having constant moments of inertia (various ratios v between 0.1 and 10 have been considered). The computations show that,

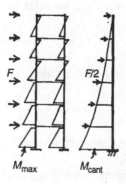

M_{max} M_{cant}

Figure 1.6

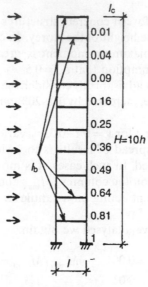

Figure 1.7

if we denote

$$v = k_b/k_{c_{max}} \tag{1.5}$$

then we may use the curves shown in Figure 1.5.

- The height of the ground floor (h_G) is greater than the height of the remaining floors (h): see Figure 1.8. The moment of inertia is assumed constant (I_c), so that $k_{c_G} = I_c/h_G < k_c = I_c/h$. Ground-floor heights $h_G = 1.5\,h$–$2\,h$ have been taken into account.

Ratios v between 0·1 and 10 have been checked.

The results show that we may use the curves of Figure 1.5 on condition that

for ε_G, we refer to

$$v_G = \frac{k_b}{k_{c_G}} = \frac{I_b/l}{I_{c_G}/h_G} \tag{1.6}$$

for ε_2, we refer to

$$v = \frac{k_b}{k_c} = \frac{I_b/l}{I_c/h} \tag{1.7}$$

More accurate results may be obtained in this latter case by using the continuum approach (see section 1.2.3).

In the case of variable heights and moments of inertia, the results obtained by the ZMP procedure are only reliable for the ground floor.

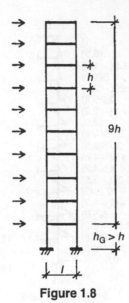

Figure 1.8

□ **Numerical example 1.1**

Consider the uniform one-bay, 10-storey frame shown in Figure 1.9, loaded by horizontal identical concentrated loads $F = 1\,\text{kN}$. $I_c = 1$, i.e. $k_c = 0\cdot333$. Assume $I_b = 1\cdot333$, i.e. $k_b = 1\cdot333/4 = 0\cdot333$; $v = k_b/k_c = 1$.

The accurate moments are shown in Figure 1.9b (within the brackets). To compute the same moments approximately, we use the curves given in Figure 1.5: $v = 1$, $\varepsilon_G = 0\cdot66$, $\varepsilon_2 = 0\cdot53$, $\varepsilon_m = 0\cdot50$. The corresponding moments are shown in Figure 1.9b (outside the brackets).

For instance, at the ground floor:

$$M_{\text{bott}} = \frac{1}{2}\sum_1^{10} F\cdot\varepsilon_G\cdot h_G = \frac{10}{2}\times 0\cdot66\times 3\cdot0 = 9.9\,\text{kN\,m}\quad\text{(accurate: }9\cdot5\,\text{kN\,m).}$$

$$M_{\text{top}} = -\frac{1}{2}\sum_1^{10} F\cdot(1-\varepsilon_G)\cdot h_G = -\left(\frac{10}{2}\right)\times 0\cdot34\times 3\cdot0 = -5\cdot1\,\text{kN\,m}$$

$$\text{(accurate: }-5\cdot5\,\text{kN\,m).}$$

By assuming a different ratio, $v = k_b/k_c = 0\cdot1$, the accurate moments are as shown in Figure 1.9c. From the curves in Figure 1.5, $\varepsilon_G = 1\cdot30, \varepsilon_2 = 0\cdot88$; ε_m is uncertain. The approximate moments are shown outside the brackets.

At the ground floor:

$$M = \frac{10}{2}\times 1\cdot30\times 3\cdot0 = 19\cdot5\,\text{kN\,m}\quad\text{(accurate: }18\cdot9\,\text{kN\,m)}$$

$$M = -\frac{10}{2}\times(1-1\cdot3)\times 3\cdot0 = 4\cdot5\,\text{kN\,m}\quad\text{(accurate: }4\cdot5\,\text{kN\,m)}$$

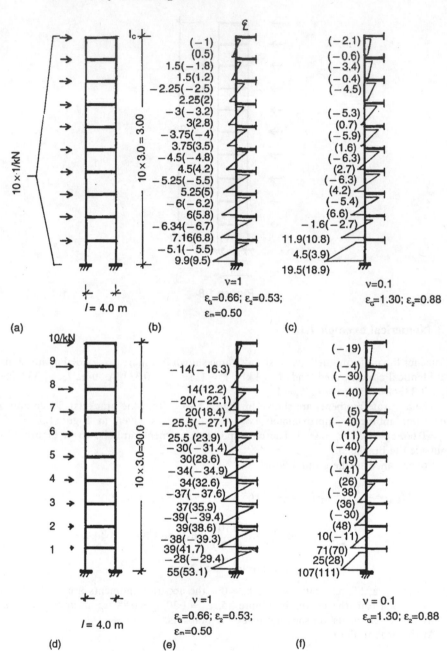

Figure 1.9

If we refer to the **cantilever moment**:

$$M_{cant} = \frac{10}{2} \times 3.0 \times (1 + 2 + \cdots + 10) = 82.5$$

yielding:

$$\frac{M_{max}}{M_{cant}} = \frac{18.9}{82.5} = 0.23$$

Similar cases are shown in Figure 1.9d–f, but for an inverted triangular load: the same ratios v as considered in the first case (uniformly distributed loads) have been assumed.

Referring to the cantilever moment:

$$M_{cant} = \left(\frac{3.0}{2}\right) \times (1 \cdot 1 + 2 \cdot 2 + \cdots + 10 \cdot 10) = 577.5$$

yielding:

$$\frac{M_{max}}{M_{cant}} = \frac{111}{577.5} = 0.19 \qquad \square$$

□ **Numerical example 1.2**

The frame shown in Figure 1.10 has variable heights, as well as variable moments of inertia. The approximate moments for the ground floor only are computed as follows.

$$v_G = \frac{k_{bG}}{k_{cG}} = \frac{1.2/3.0}{1.5/4.5} = 1.2$$

From Figure 1.5: $\varepsilon_G = 0.63$

$$M_{bott} = \frac{10}{2} \times 0.63 \times 4.5 = 14.2\,\text{kN m} \quad \text{(accurate: } 14.1\text{)}$$

$$M_{top} = -\frac{10}{2} \times 0.37 \times 4.5 = -8.3\,\text{kN m} \quad \text{(accurate: } -8.4\text{)} \qquad \square$$

1.2.3 CONTINUUM APPROACH

This procedure is based on replacing the beams of the one-bay, multi-storey building frame by a continuous medium. It yields satisfactory results provided the frame is uniform, or nearly uniform: i.e. identical moments of inertia of columns, I_c; identical moments of inertia of beams, I_b (except the top slab, where $I'_b = I_b/2$); and not very different heights of columns ($h_i/h_j = 2/3...3/2$).

Let us consider a completely uniform frame subjected to a laterally distributed load of intensity p_x (Figure 1.11a). The beams are replaced by a continuous medium formed by an infinite number of very thin horizontal laminae at

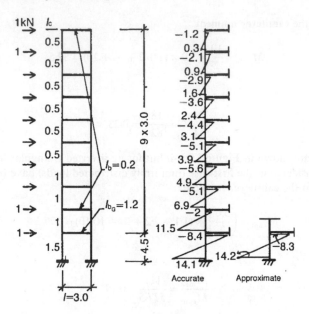

Figure 1.10

distances dx, having the moment of inertia (Figure 1.11b):

$$I'_b = \frac{I_b dx}{h} \tag{1.8}$$

By using the flexibility method, we may formulate the differential equation of compatibility of displacements and subsequently determine the couples C acting on the columns at each floor (Figure 1.11c).

For a uniformly distributed load:

$$p = \frac{\sum F}{H} \tag{1.9}$$

we obtain (Csonka, 1962a):

$$\left.\begin{array}{c} C_i = \left(\dfrac{p}{\alpha^2}\right) \cdot \sinh\left(\dfrac{\alpha h}{2}\right) \cdot [\theta \cdot \cosh(\alpha x_i) - \sinh(\alpha x_i)] + \dfrac{phx_i}{2} \\[2mm] (i = 1, 2, \ldots, n-1) \\[2mm] C_{\text{top}} = \left(\dfrac{p}{2\alpha^2}\right) \cdot \left[\theta \sinh\left(\dfrac{\alpha h}{2}\right) - \cosh\left(\dfrac{\alpha h}{2}\right) + 1\right] + \dfrac{ph^2}{16} \end{array}\right\} \tag{1.10}$$

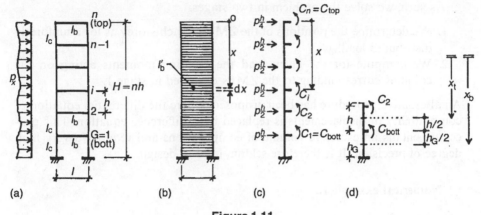

Figure 1.11

where:

$$\alpha = \sqrt{\left(\frac{6I_b}{I_c \cdot h \cdot l}\right)}\ \mathrm{m}^{-1}; \qquad \theta = \frac{\sinh(\alpha H) - \alpha H}{\cosh(\alpha H)}\ \text{(for } H > 20,\ \alpha = 1\text{)}. \quad (1.10)$$

We note that the coordinates x_i originate at the top.

Each column is now acted upon by the couples C_i and the given loads $p/2\,\mathrm{kN\,m^{-1}}$, i.e. $ph/2\,\mathrm{kN}$ at each joint, except the top joint, where $ph/4$ and C_{top} are acting (Figure 1.11c). The problem is statically determinate, and we may compute the bending moments:

$$M_i = \frac{M_{p_i}}{2} + M_{c_i} \qquad\qquad (1.11)$$

$$\text{where } M_{p_i} = \frac{p x_i^2}{2}; \qquad M_{c_i} = \sum_1^i C \qquad\qquad (1.11)$$

In the case of a different height at ground floor ($h_G \neq h$), the corresponding couple to be considered is (Figure 1.11d)

$$C_{\mathrm{bott}} = \left(\frac{p}{2\alpha^2}\right) \cdot \{\theta[\sinh(\alpha x_B) - \sinh(\alpha x_T)] - (\cosh(\alpha x_B) - \cosh(\alpha x_T)\}$$

$$+ \frac{p(x_B^2 - x_T^2)}{4} \qquad\qquad (1.12)$$

The results in this case are not very accurate.

In fact, we may use the results obtained for the uniformly distributed loads p (namely, the positions of the ZMPs) for an approximate analysis of the frame subjected to any lateral distributed load by assuming that the ZMPs are the same.

As such, we solve the problem in two stages.

1. We determine the positions of the ZMP at each storey as for a uniformly distributed load.
2. We compute for the given load the bending moments acting on the columns corresponding to the ZMPs obtained in stage 1.

An alternative procedure has been proposed, where the differential equation of compatibility of displacements is replaced by a difference equation. This procedure entails an excessive volume of computations and requires a very high degree of precision. It is therefore seldom used in design.

□ **Numerical example 1.3**

Let us refer to the frame shown in Figure 1.12 (a nearly uniform frame): $n = 10$; $l = 3 \cdot 0$ m; $h = 3 \cdot 0$ m; $h_G = 4 \cdot 5$ m; $H = 31 \cdot 5$ m; $I_b/I_c = 1 \cdot 5$.

$$p = \frac{1 \cdot 0}{3 \cdot 0} = 0 \cdot 333 \, \text{kN m}^{-1} \quad (\text{or: } F_i = 1 \, \text{kN}; \ F_{\text{top}} = 0 \cdot 5 \, \text{kN}; \ F_{\text{bott}} = 1 \cdot 25 \, \text{kN}).$$

$$\alpha = \sqrt{\left(\frac{6 \times 1 \cdot 5}{1 \times 3 \times 2}\right)} = 1; \quad \theta = 1$$

$$C_i = \left(\frac{p}{\alpha^2}\right) \cdot \sinh\left(\frac{\alpha}{2}\right) \cdot [\theta \cdot \cosh(\alpha x_i) - \sinh(\alpha x_i)] + \frac{phx_i}{2} = \cdots$$

$$= 0 \cdot 709 \cdot [\cosh(\alpha x_i) - \sinh(\alpha x_i)] + 0 \cdot 5 \, x_i$$

$i = 2$	$x_i = 24$ m	$C_i = 11 \cdot 99$ kN m
3	21	10·49
4	18	8·99
5	15	7·50
6	12	6·00
7	9	4·50
8	6	3·00
9	3	1·53

$$C_{\text{top}} = \left(\frac{p}{2\alpha^2}\right) \cdot \left[\theta \sinh\left(\frac{\alpha h}{2}\right) - \cosh(\alpha h/2) + 1\right] + \frac{p \cdot h^2}{16}$$

$$= \left(\frac{0 \cdot 333}{2 \times 1^2}\right) \cdot (1 \times 2 \cdot 1293 - 2 \cdot 3524 + 1) + \frac{0 \cdot 333 \times 3^2}{16} = 0 \cdot 32 \, \text{kN m}$$

$$C_{\text{bott}} = \left(\frac{p}{2\alpha^2}\right) \cdot \{\theta \cdot [\sinh(\alpha x_B) - \sinh(\alpha x_T)] - (\cosh(\alpha x_B) - \cosh(\alpha x_T)\}$$

$$+ \frac{p(x_B^2 - x_T^2)}{4}$$

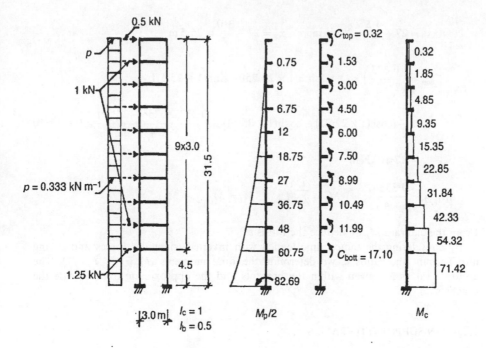

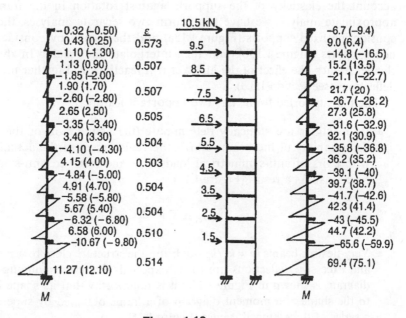

Figure 1.12

$$x_B = 27 + \frac{4 \cdot 5}{2} = 29 \cdot 25 \, \text{m}; \qquad x_T = 27 - \frac{3 \cdot 0}{2} = 25 \cdot 5 \, \text{m}$$

$$C_{\text{bott}} = \left(\frac{0 \cdot 333}{2 \times 1^2}\right) \times \{1 \times [\sinh(1 \times 29 \cdot 25) - \sinh(1 \times 25 \cdot 5)]$$

$$- (\cosh(1 \times 29 \cdot 25) - \cosh(1 \times 25 \cdot 5)\} + \frac{0 \cdot 333(29 \cdot 25^2 - 25 \cdot 5^2)}{4} \cong 0 + 17 \cdot 10$$

$$= 17 \cdot 10 \, \text{kN m}$$

$$M_{(P_i/2)} = \frac{0 \cdot 333 \, x_i^2}{4}; \qquad M_{c_i} = \sum_1^i M_c; \qquad M_i = M_{(P_i/2)} + M_{c_i}.$$

From the diagram M_i we deduce the ratios ε.

By considering the same frame loaded with inverted triangular forces and taking into account the ratios ε, we determine the final moments ($M_i = \sum_1^i F\varepsilon h/2$). The accurate values are given within the brackets and the approximate ones outside the brackets. □

1.2.4 PIN-SUPPORTED FRAMES

Multi-storey building frames with pinned supports are seldom used in design, especially if the columns are of reinforced concrete. The actual supports are elastic, but designers usually consider them as fixed. If we want to take into account the elasticity of the supports against rotation in the frame of an approximate analysis we have to perform two separate analyses, the first by considering fixed supports (Figure 1.13a) and the second by considering pinned supports (Figure 1.13b), and then interpolate the results. In the case of slender frames, the effect of the high vertical reactions will further increase the deflections (see section 1.3.3).

The pin-supported frame has two important features.

• Its reactions are statically determinate (the assumption of the axial indeformability of the beams allows the replacement of the actual loads F with pairs of anti-symmetrical loads $F/2$ and leads to anti-symmetrical reactions). As a result (Figure 1.14):

$$M_G = \sum F \cdot \frac{h_G}{2}$$

• Very flexible beams ($v = k_b/k_c \to 0$) bring the structure close to a mechanism and excessive deflections are to be expected. The corresponding moment diagram is shown in Figure 1.15a; it is noteworthy that its shape is similar to the shape of a moment diagram of a frame obtained by superimposing a series of three-hinged frames (Figure 1.15b).

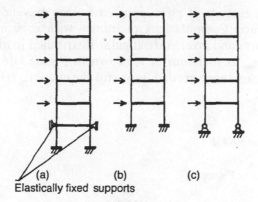

(a)   (b)   (c)

Elastically fixed supports

Figure 1.13

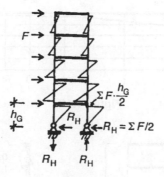

Figure 1.14

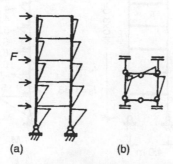

(a)   (b)

Figure 1.15

The approximate analysis of pin-supported frames may be performed by the ZMP procedure. Computations of frames with 6, 10 and 15 storeys subjected to uniform and inverted triangular distributed loads have yielded curves of ε_2 and ε_m as functions of v, shown in Figure 1.16. The curve ε_2 refers to the columns above ground floor and the curve ε_m refers to floors at mid-height.

☐ **Numerical example 1.4**

For the frame shown in Figure 1.17:

$$v = \frac{k_b}{k_c} = \frac{1 \cdot 33/4 \cdot 0}{1/3.0'} = 1$$

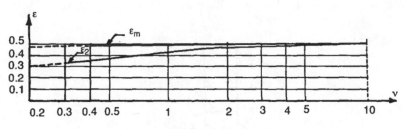

Figure 1.16

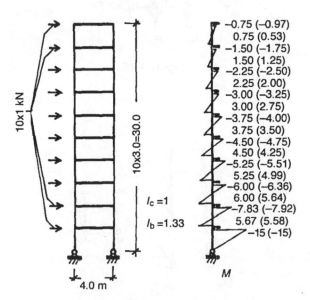

Figure 1.17

From the curves of Figure 1.16: $\varepsilon_2 \cong 0\cdot42$: $\varepsilon_m \cong 0\cdot50$
$M_{2_{bott}} = 9 \times 3\cdot0 \times 0\cdot42/2 = 5\cdot67\,\mathrm{kN\,m}$ (accurate: 5·58)
$M_{4_{bott}} = 7 \times 3\cdot0 \times 0\cdot5/2 = 5\cdot25\,\mathrm{kN\,m}$ (accurate: 4·99).   □

1.2.5 APPROXIMATE ASSESSMENT OF DEFLECTIONS

It is of practical interest to assess the horizontal deflections of multi-storey buildings subjected to lateral loads, with special reference to the following:

- the general deformed shape, in order to compute the dynamic characteristics of the structure and to identify the storeys most exposed to excessive drift;
- the maximum deflection (at the top floor), in order to check the general flexibility of the structure;
- the maximum storey drifts (usually limited by codes: e.g. the model code CEB-85).

The deformed shape of the structure depends on the ratio $v = k_b/k_c$ and, to a lesser extent, on the type of loading.

Typical deformed shapes are shown in Figure 1.18 for three different ranges of the ratio v, related to uniform fixed frames subjected to distributed loads. We note that in the usual range ($v = 0\cdot1-5$) the deformed shapes are rather close to straight lines (the maximum deviation is $\Delta u/u_{max} < 20\%$; this is in agreement with the assumption adopted in most earthquake regulations that, for regular structures, a straight line may be used in assessing the distribution of seismic loads corresponding to the fundamental mode of vibration (see e.g. CEB-85).

Typical deformed shapes for uniform pinned frames are shown in Figure 1.19. We note that, for very flexible beams ($v < 0\cdot01$), the deformed shape tends to a straight line. For $v \to 0$, it indicates the vicinity of instability: the structure becomes a mechanism and the columns rotate without deformations. With stiff beams ($v > 2$), the deformed shape tends to two straight lines joining somewhere between the first and the second floor. In the usual range ($v = 0.1-5$),

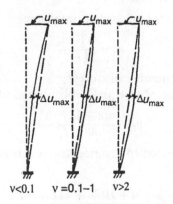

$v<0.1$ $v=0.1-1$ $v>2$

Figure 1.18

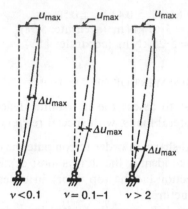

Figure 1.19

the deviation from the straight line is much greater than the deviation of fix-supported frames; the straight-line assumption in this instance is no longer justified.

The deflections are very sensitive to changes in geometry and rigidity, and therefore it is rather risky to make an approximate assessment; we therefore only give indications as to the order of magnitude.

We shall represent u_{max} in the form

$$u_{max} = \mu \cdot u_R \tag{1.13}$$

where we define by u_R the maximum deflection computed on the assumption that the beams are rigid ($v = k_b/k_c \to \infty$) and by u_{max} the actual maximum deflection (Figure 1.20).

Note that all computations performed in this chapter are based on the assumption that the effect of axial forces and shear forces on the deformations is neglected (see section 1.1).

Figure 1.20 presents four cases:

- a uniform frame subjected to a concentrated force F at the top (Figure 1.20a):

$$u_R = \frac{n F h^3}{24 E I_c} \tag{1.14}$$

    where n denotes the number of storeys;
- the same frame subjected to equal concentrated loads (Figure 1.20b):

$$u_R = \frac{n(n+1) F \cdot h^3}{48 E I_c} \tag{1.15}$$

- the same frame subjected to inverted triangular concentrated loads (Figure 1.20c):

$$u_R = \frac{(n+1)(2n+1) F_{max} \cdot h^3}{144 E I_c} \tag{1.16}$$

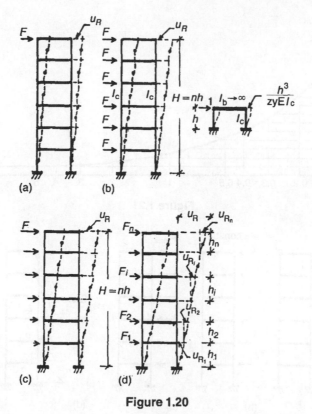

Figure 1.20

- a general case: a frame where I_{b_i}, I_{c_i} and F_i are variable (Figure 1.20d):

$$u_{max} = \sum_{i=1}^{n} \mu_i \cdot u_{R_i} \qquad (1.17)$$

where

$$u_{R_i} = \left(\frac{1}{24}\right) \cdot \left[(F_1 + F_2 + \cdots + F_n)\frac{h_1^3}{EI_{c_1}} \right.$$

$$\left. + (F_2 + F_3 + \cdots + F_n)\frac{h_2^3}{EI_{c_2}} + \cdots + \frac{F_n h_n^3}{EI_{c_n}} \right]$$

μ_i: according to the ratio $v_i = k_{b_i}/k_{c_i}$.

An average curve of μ versus v is displayed in Figure 1.21, which may be used for any laterally distributed load.

For ratios $v < 0.1$, we have to compare the maximum actual deflection u_{max} with the maximum cantilever deflection u_{cant}; for constant moments of inertia of cantilevers:

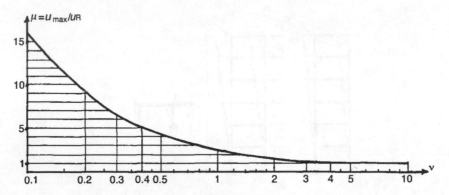

Figure 1.21

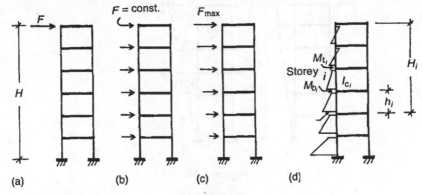

Figure 1.22

- a concentrated force at the top (Figure 1.22a):

$$u_{cant} = \frac{F H^3}{6 E I_c}$$

(1.18)

- equal concentrated loads F (Figure 1.22b):

$$u_{cant} = \frac{n F H^3}{16 E I_c}$$

(1.19)

- inverted triangular loads (Figure 1.22c):

$$u_{cant} = \frac{11 (n+1) F_{max} H^4}{240 E I_c}$$

(1.20)

Numerical examples indicate the following ratios:

$$v = 0 \cdot 001 \qquad u_{max}/u_{cant} \cong \quad 0 \cdot 8 - 0 \cdot 9$$
$$0 \cdot 01 \qquad\qquad\qquad 0 \cdot 3 - 0 \cdot 4$$
$$0 \cdot 1 \qquad\qquad\qquad 0 \cdot 04 - 0 \cdot 08$$

(1.21)

In the case of nearly uniform multi-storey frames, we may determine the columns moment diagram by using the continuum approach and then compute the maximum deflection (Figure 1.22d):

$$u_{max} = \sum_{1}^{n} \left(\frac{h_i}{2 E I_{c_i}} \right) \cdot \left[M_{b_i} \left(H_i - \frac{h_i}{3} \right) + M_{t_i} \left(H_i - \frac{2h_i}{3} \right) \right] \qquad (1.22)$$

As mentioned above, we have to check the maximum drift. For regular uniform frames, the deformed shape is almost a straight line, and consequently we may assume

$$\frac{\Delta u}{h} \cong \frac{u_{max}}{H} \qquad (1.23)$$

As may be observed in Figure 1.18, the location of the maximum drift depends on the ratio $v = k_b/k_c$: for flexible beams ($v < 0.1$), the maximum drift occurs at the top floor, while for stiff beams ($v > 2$) it occurs at ground floor. When $h_G > h$, the maximum drift occurs usually at the ground floor. In this latter case:

$$\frac{u_G}{h_G} \cong (3\varepsilon_G - 1) \frac{(\sum F) h_G^2}{12 E I_G} \qquad (1.24)$$

ε_G will be taken from the curves given in Figure 1.5.

The maximum deflections of pin-supported frames (u_p) may be assessed as a percentage of the corresponding maximum deflection of the fixed end frame (u_p/u_f).

Referring to uniform frames with 6–15 storeys subjected to laterally distributed loads, the ratio u_p/u_f varies as follows:

$$v = 0{\cdot}001 \qquad u_p/u_f = \qquad 7{-}10$$
$$0{\cdot}01 \qquad\qquad\qquad 2{-}3$$
$$0{\cdot}1 \qquad\qquad\qquad 1{\cdot}4{-}1{\cdot}6$$

For the usual range $v = 0{\cdot}1{-}5$, we may consider

$$\frac{u_P}{u_f} \cong 1{\cdot}5 \qquad (1.26)$$

As mentioned above, in the case of pin-supported frames with very flexible beams the deflections tend to infinity (the structure becomes unstable).

☐ **Numerical example 1.5**

Consider the uniform, fix-supported frame shown in Figure 1.23a:

$$I_G = I_c; \qquad \frac{I_b}{I_c} = 1{\cdot}33; \qquad v = k_b/k_c = 1; \qquad v_G = k_b/k_c = 1{\cdot}5$$

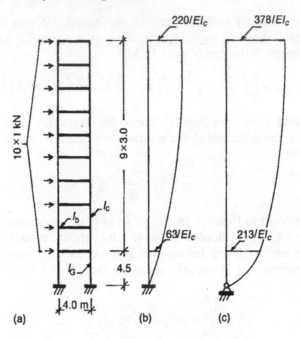

Figure 1.23

From the curve $\mu = u_{\text{max}}/u_R$ (Figure 1.21): $\mu \cong 2\cdot7$; $\mu_G \cong 2\cdot2$ According to equation (1.17):

$$u_{\text{max}} = \mu_G \cdot u_{RG} + \mu u_{R2.n}; \qquad u_{RG} = \frac{n F h^3}{24 E I_c} = \frac{10 \times 1 \times 4\cdot5^3}{24 E I_c} = \frac{37\cdot97}{E I_c}$$

Since

$$\frac{n(n+1)}{48} - \frac{n}{24} = \frac{n(n-1)}{48}$$

$$u_{R2.n} = \frac{n(n-1) F h^3}{48 E I_c} = \frac{10 \times 9 \times 1 \times 3^3}{48 E I_c} = \frac{50\cdot62}{E I_c}$$

$$u_{\text{max}} = \frac{(2\cdot2 \times 37\cdot97 + 2\cdot7 \times 50\cdot62)}{E I_c} = \frac{220}{E I_c}$$

(accurate: $210/E I_c$).

The deformed shape is shown in Figure 1.23b. The maximum drift is to be expected at the ground floor. According to equation (1.24):

$$\frac{u_G}{h_G} = \frac{(3\varepsilon_G - 1)(\Sigma F) h_G^2}{12 E I_G}$$

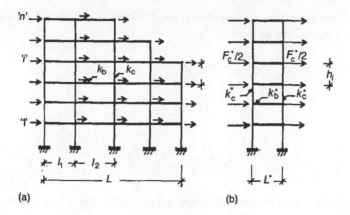

Figure 1.24

From the curve ε/v (Figure 1.5), for $v_G = 1\cdot5$, we obtain $\varepsilon_G \cong 0\cdot61$.

$$\frac{\Delta u_G}{h_G} \cong \frac{(3 \times 0.61 - 1) \times 10 \times 4.5^2}{12\,E\,I_c} = \frac{14.0}{E\,I_c}$$

(accurate: $5\cdot6/4\cdot5\,E\,I_c = 12\cdot4/E\,I_c$).

By considering the same frame, but pin-supported, according to equation (1.26):

$$\frac{u_p}{u_f} \cong 1\cdot5; \qquad u_p \cong \frac{1.5 \times 220}{E\,I_c} = \frac{330}{E\,I_c} \quad \text{(accurate: } 378/E\,I_c\text{)}$$

The deformed shape is shown in Figure 1.23c. The actual drift at ground floor is much greater than the precedent: $u_G/h_G = 213/4\cdot5E\,I_c = 47\cdot3/E\,I_c$ as against: $63/4\cdot5E\,I_c = 14/E\,I_c$ (for the fix-supported frame). □

For computation of u_{max}, in the case of variable heights and moments of inertia, see numerical example 1.6 in section 1.3.2.

1.3 The substitute (equivalent) frame method

1.3.1 GENERAL APPROACH

The method of the substitute (equivalent) frame is an approximate method of analysing a multi-storey, multi-bay frame by replacing it by a multi-storey, one-bay frame.

In the following we adopt the same assumptions that have been formulated in section 1.1.

Let us consider the building frame shown in Figure 1.24a. The substitute frame is defined in Figure 1.24b. Its storey heights are the same as the storey heights of the given frame, but the span L^* may be chosen arbitrarily. Its moments of inertia are chosen so that

$$\left.\begin{array}{c} \sum k_c^* = 2k_c^* = \sum k_c; \qquad k_c^* = \dfrac{\sum k_c}{2} \\[2mm] k_b^* = \sum k_b \end{array}\right\} \qquad (1.27)$$

where

$$k_c = \frac{I_c}{h}; \qquad k_b = \frac{I_b}{l}$$

The total loads acting at each floor are the same for the given and the substitute frames:

$$F^* = \sum F \qquad (1.28)$$

Numerous computations have shown that the substitute frame exhibits two important features:

$$u^* \cong u \qquad (1.29)$$

where u is the deflection of the given frame and u^* is the deflection of the substitute frame, and

$$\sum M_c^* = 2M_c^* \cong \sum M_c; \qquad M_c^* \cong \frac{\sum M_c}{2} \qquad (1.30)$$

where M_c is the bending moment in the columns of the given frame and M_c^* is the bending moment in the columns of the substitute frame. We shall admit the distribution of the total moment $\sum M_c^* = 2M_c^*$ at any storey among the columns of the given frame, in proportion to their moments of inertia:

$$M_{c_j} = 2M_c^* \frac{I_{c_j}}{\sum I_{c_j}} \qquad (1.31)$$

The validity of this assumption is considered in the following.

The method of the substitute frame is an approximate one. In a single case, it becomes an accurate method (within the limits of the assumptions defined in section 1.1), if the given frame is a 'proportioned' one (see Figure 1.25). This frame may be considered as consisting of a number of identical one-bay frames, each one loaded with the same loads (Figure 1.25a, b). As their deformed shapes are identical, when connected in order to obtain the original multi-bay frame, no contact forces develop in these connections. Consequently:

- the deformed shape of the original multi-storey frame is identical to the deformed shape of the elementary one-bay frame;

- the sum of the bending moments in the columns at any level of the given frame is equal to the sum of the corresponding bending moments in all the elementary one-bay frames.

We may now replace the elementary frames by the substitute frame shown in Figure 1.25c, where

$$\sum k_c^* = 2 k_c^* = \frac{2 I_c^*}{h} = \frac{\sum I_c}{h}$$

or

$$2 I_c^* = \sum I_c; \qquad k_b^* = \frac{I_b^*}{L^*} = \sum k_b = \frac{\sum I_b}{l}$$

loaded with $F^* = \sum F$. This will lead to the accurate relationships

$$u^* = u; \qquad M_c^* = \frac{\sum M_c}{2}$$

Note that the effect of axial and shear forces on the deformations has been neglected (see section 1.1).

Obviously, the 'proportioned frame' is a theoretical concept, one that is most unlikely to be encountered in actual design.

1.3.2 REMARKS DEALING WITH THE APPLICATION OF THE SUBSTITUTE FRAME METHOD

The distribution of the total moment $\sum M_c = 2 M_c^*$ among the columns of the given frame according to their moments of inertia is accurate only in two cases: when the beams are very stiff ($v = k_b/k_c \to \infty$) or when they are very flexible ($v \to 0$). In the intermediate range (most of the actual cases), the relative rigidities of the internal columns are higher and the relative rigidities of the end columns are lower than those yielded by the ratios k_c^{int}/k_c^{side}; as we see from Figure 1.26 the end column deformation involves a beam segment $\sim l/2$, while the deformations of the internal columns involve a beam segment $\sim 2l/2 = l$. This property is less significant at the ground floor, where the bottom supports are identical for both types of columns.

From the numerical examples, we conclude that:

- we may neglect this effect for $v < 0.05$ and $v > 10$;
- we have to take it into account in the domain $0.05 < v < 10$.

Several procedures have been proposed in the technical literature, aimed at improving the accuracy of distribution of the total moment $2 M_c^* = \sum M_c$ between the columns [see e.g. R. Smith, cited by Rutenberg (1966), Muto (1974)], but these techniques require excessive computation, which is incompatible with the purpose of using approximate methods.

We propose therefore:

- To increase the moments acting on the internal columns, computed according to equation (1.31), by multiplying them by the following **corrective coefficients**:

for the ground floor: $M_c^{\text{corrected}} = (1 \cdot 1 \ldots 1 \cdot 2) M_c$

for the other floors: $M_c^{\text{corrected}} = (1 \cdot 2 \ldots 1 \cdot 3) M$

- to leave the moments acting on the end columns, computed according to equation (1.31), undiminished, so as to remain on the safe side.

After distributing the total moment $2 M_c^* = \sum M_c$ among columns, we may deduct the end moments acting on the beams.

For an end column (Figure 1.27), the equation of equilibrium suffices:

$$M_b = M_c' + M_c'' \tag{1.32}$$

For an internal column, we shall distribute the sum of moments acting on the column in proportion to the relative rigidities of the beams:

$$M_b^L = \frac{(M_c' + M_c'') \, k_b^L}{k_b^L + k_b^R}$$

$$M_b^R = \frac{(M_c' + M_c'') \, k_b^R}{k_b^L + k_b^R} \tag{1.33}$$

In assessing the maximum vertical reactions due to horizontal loading, we note that in most cases the reactions occurring at the extremities are much greater than the intermediate ones (Figure 1.28). Consequently, so as to remain on the safe side:

$$R_v^{\text{max}} = M_{\text{TOT}}/L = \frac{M_{\text{ext}} - \sum M_f}{L} = \frac{M_{\text{ext}} - F_{\text{TOT}} \, \varepsilon_G \, h_G}{L} \tag{1.34}$$

where M_{ext} is the moment of the horizontal loads with respect to any point lying along the support line (the **overturning moment**); $\sum M_f$ is the sum of the support moments.

We note that the vertical reactions due to horizontal loads are usually smaller than the reactions due to vertical loads, especially for relatively low frames. Bearing in mind that currently accepted codes allow for an increase in the stresses when horizontal loads are involved, we often can neglect the additional vertical reactions due to horizontal loads, but it is desirable to take these additional reactions into account for frames where $H > L$.

As already emphasized, we may arbitrarily choose the span of the substitute frame (L^*). To obtain the vertical reactions directly from the analysis of the substitute frame, we shall choose a span equal to the total width of the given multi-bay frame ($L^* = L$).

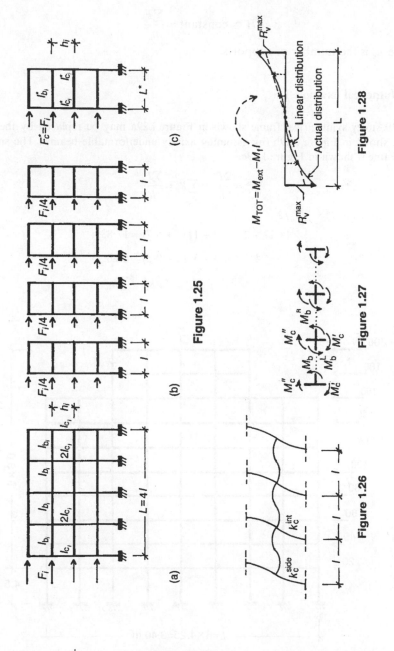

Figure 1.25

Figure 1.26

Figure 1.27

Figure 1.28

The horizontal reactions are more or less uniformly distributed between the supports:

$$H \cong \text{constant} = \frac{\Sigma F}{n_s} \tag{1.35}$$

where n_s is the number of supports.

□ Numerical example 1.6

The 10-storey symmetrical frame shown in Figure 1.29a may be replaced by the half frame shown in Figure 1.29b (by assuming axially underformable beams). The substitute frame is shown in Figure 1.29c.

$$k_c^* = \frac{2 I_c^*}{h} = \Sigma k_c = \frac{\Sigma I_c}{h}$$

As $h = \text{constant}$, $I_c^* = \Sigma I_c / 2$.

$$\tfrac{1}{2} \times (4 \cdot 17 + 2 \times 7 \cdot 2 + 11 \cdot 4 + 5 \cdot 7) = 17 \cdot 835;$$

$$\tfrac{1}{2}(1 \cdot 6 + 3 \times 3 \cdot 12 + 0 \cdot 5) = 6 \cdot 26; \dots$$

$$k_b^* = \Sigma k_b$$

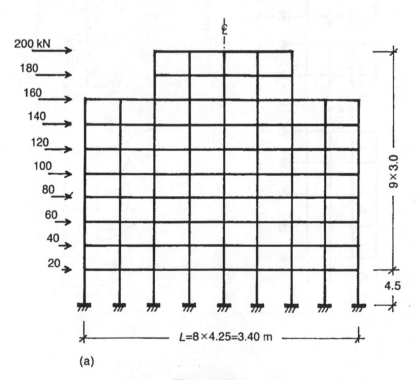

(a)

Figure 1.29

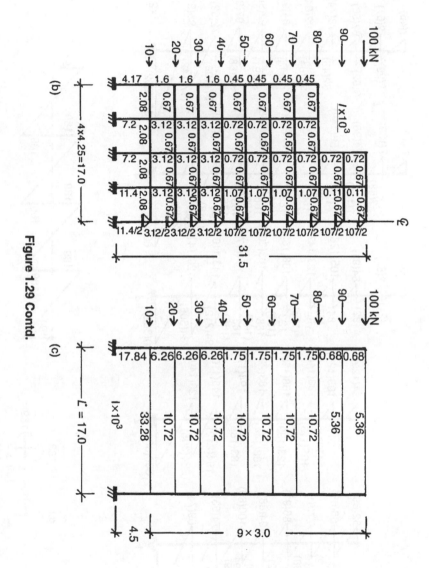

(b)

(c)

Figure 1.29 Contd.

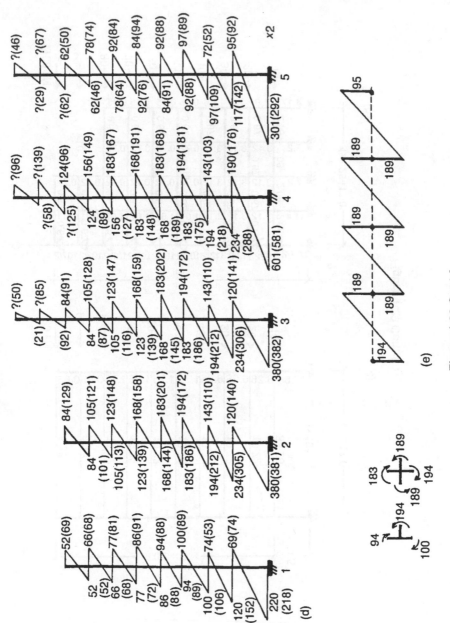

Figure 1.29 Contd.

We choose $L^* = L = 17.00$.

$$\frac{I_b^*}{17} = \frac{4 \times 2.08}{4.25}; \quad I_b^* = 33.28; \quad \frac{I_b^*}{17} = \frac{4 \times 0.67}{4.25}; \quad I_b^* = 10.72\ldots$$

Bending moments in columns (Figure 1.29d):

$$\text{At ground floor: } v_G = \frac{33.28/17}{17.835/4.5} \cong 0.5$$

From the curves given in Figure 1.5: $\varepsilon_G = 0.76$.

$$M_{bott}^* = \left(\frac{\sum F}{2}\right)\varepsilon_G h_G = \left(\frac{550}{2}\right) \times 0.76 \times 4.5 = 941$$

$$M_{top}^* = -\left(\frac{\sum F}{2}\right)(1 - \varepsilon_G) h_G = -\left(\frac{550}{2}\right) \times 0.24 \times 4.5 = -297$$

We distribute the total moments $\sum M_c$ between the columns of the given frame according to their moments of inertia (Figure 1.29d):

$$\text{Column 1: } M_{bott} = \frac{2 \times 941 \times 4.17}{35.67} \cong 220$$

$$M_{top} = \frac{-2 \times 297 \times 4.17}{35.67} = -69.4$$

$$\text{Column 2: } M_{bott} = \frac{2 \times 941 \times 7.2}{35.67} \cong 380$$

$$M_{top} \cong -120$$
$$\ldots$$

The accurate results are shown in Figure 1.29d within the brackets, and the approximate ones outside the brackets. We have to multiply the approximate moments of the internal columns by 1.1–1.2.

As mentioned abvoe, the results obtained for the second storey are much less reliable:

$$v_2 = \frac{10.72/17}{6.26/3} = 0.3$$

Curve of Figure 1.5: $\varepsilon_2 \cong 0.58$.

$$M_{bott} = \frac{540}{2} \times 0.58 \times 3.0 = 470$$

$$M_{top} = -\left(\frac{540}{2}\right) \times 0.42 \times 3.0 = -288$$

$$\text{Column 1: } M_{bott} = \frac{2 \times 470 \times 1.6}{12.52} = 120$$

$$M_{top} = \frac{-2 \times 288 \times 1.6}{12.52} \cong -74$$

At the intermediate floors: $v = 0 \cdot 30 \ldots 1 \cdot 08$.

Curve of Figure 1.5: $\varepsilon_m \cong 0.5$ (ZMP at mid-heights of storeys).

$$M_{\text{bott}} = - M_{\text{top}} = \frac{\sum F h}{2}$$

The approximate results at the top storeys are unreliable, owing to the change in the geometrical configuration and the resulting discontinuity in rigidity. As mentioned above, they are not essential to the design.

The end moments acting on the beams of the third floor are shown in Figure 1.29e (within brackets: accurate results).

The maximum vertical reactions may be computed from Figure 1.28:

$$R_v^{\max} = \frac{M_{\text{TOT}}}{L^*} = \frac{M_{\text{ext}} - \sum M_f}{L}$$

$$L^* = 17\text{m}; \qquad M_{\text{ext}} = 10 \times 4 \cdot 5 + 20 \times 7 \cdot 5 + \cdots + 100 \times 31 \cdot 5$$

$$= 12375 \text{ kN m}$$

$$\sum M_f = 2 M_f^* = 2 \times 924 = 1848 \text{ kN m};$$

$$R_v^{\max} \cong (12375 - 1848)/17 = 619 \text{ kN}$$

(accurate: 566 kN).

The reaction on the axis of symmetry is zero:

$$R_v^{\text{sym}} = 619 - 619 = 0$$

The maximum horizontal deflection will be computed according to equation (1.17):

$$u_{\max} = \sum \mu_i \cdot u_{R_i}; \qquad u_{R_i} = \frac{\sum F_i h_i^3}{24 E I_{c_i}}$$

μ_i is a coefficient taken from the curve shown in Figure 1.21.

Ground floor:

$$v = \frac{k_b}{k_c} = \frac{33 \cdot 28/17}{17 \cdot 835/4 \cdot 5} \cong 0 \cdot 5 \ldots \mu \cong 4 \cdot 3$$

Storeys 2, 3, 4:

$$v = \frac{k_b}{k_c} = \frac{10 \cdot 72/17}{6 \cdot 26/3} \cong 0 \cdot 3 \ldots \mu \cong 6 \cdot 5$$

Storeys 5, 6, 7, 8:

$$v = \frac{k_b}{k_c} = \frac{10 \cdot 72/17}{1 \cdot 75/3} \cong 1 \cdot 1 \ldots \mu \cong 2 \cdot 5$$

Storeys 9, 10:

$$v = \frac{k_b}{k_c} = \frac{5 \cdot 36/17}{0 \cdot 68/3} \cong 1 \cdot 4 \ldots \mu \cong 2 \cdot 2$$

$$\frac{E\,u_{max}}{1000} = 550 \times 4\cdot5^3 \times \frac{4\cdot3}{(24 \times 17\cdot84)} + (540 + 520 + 490) \times 3^3$$

$$\times \frac{6\cdot5}{24 \times 6\cdot26} + (450 + 400 + 340 + 270) \times 3^3 (2.5)$$

$$\times \frac{1}{24 \times 1\cdot75} + (190 + 100) \times 3^3 \times \frac{2.2}{24 \times 0\cdot68} = 5715\cdot890$$

$$E = 3 \times 10^7 \text{ kN m}^{-2}$$

$$u_{max} = \frac{5\,715\,890}{3 \times 10^7} = 0.190 \text{ m}   \text{(accurate: } u_{max} = 0\cdot187 \text{ m)}.$$

The drift at ground-floor level (equation (1.24)):

$$\frac{\Delta u_G}{h_G} = \frac{(3\,\varepsilon_G - 1)\sum F\,h_G^2}{12\,E\,I_G}$$

$$v = 0.5 \dots \varepsilon_G \cong 0.76$$

$$\frac{\Delta u_G / h_G}{1000} = \frac{(3 \times 0\cdot76 - 1) \times 550 \times 4\cdot5^2}{12 \times 3 \times 10^7 \times 17\cdot84/1000} = 0\cdot00222 \text{ m}$$

(accurate: 0.00216 m).

1.3.3 EFFECT OF SOIL DEFORMABILITY

When taking soil deformability into account, we have to replace the 'perfect supports' (fixed ends or pinned supports) by springs, i.e. horizontal, vertical and sometimes rotational springs. The spring constants are determined in agreement with the given subgrade moduli of the soil; they usually vary between $20\,000 \text{ kN m}^{-3}$ (for soft soils) and $100\,000 \text{ kN m}^{-3}$ (for hard soils).

Soil deformability leads to an increase in the horizontal deflections and a corresponding decrease of the frame's rigidity compared with the same structure on perfect supports.

Results of numerical examples (Figure 1.30) suggest increases of horizontal deflections up to 30–40% for normal frames ($H/L < 2$). For slender frames ($H/L > 2$), the increase may exceed 60–70%.

The additional vertical reactions due to the overturning moment lead to an increase in the general rotation of the structure and subsequently also to an increase in the horizontal deflections.

The method of the substitute frame in structural analysis (statics, dynamics and stability) was developed in the years 1945–1970 by Kornouhov (1949), Klouček (1950), Melnikov and Braude (1952), Lightfoot (1956), Ehlers (1957), Rutenberg (1966), Murashev (1971), Williams (1977), and Allen and Darvall (1977).

During the years 1930–1940, Muto developed the procedure of approximate analysis of multi-storey frames by using the zero moment procedure (Muto, 1974).

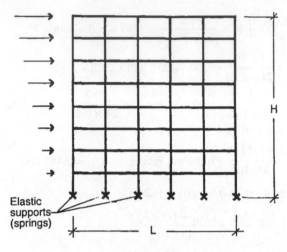

Figure 1.30

The author has proposed a computer program for the approximate analysis of multi-storey structures based on the substitute frame (Scarlat, 1986).

The main references for the analysis of one-bay frames by the continuum approach are Beck (1956), Merchant (1955), and Csonka (1962b), although the basic concepts of this method are much older (Bleich and Melan, 1927).

1.4 First screening of existing moment-resisting frame structures

1.4.1 PURPOSE OF FIRST SCREENING METHODS

In many cases, we need a rapid evaluation of the order of magnitude of a structure's capacity to resist horizontal, mainly seismic, forces. It is useful for a preliminary design, as a check of a project, and especially as a tool to check the probable seismic resistance of existing buildings.

A review of recent developments in this field, as well as a technique proposed by the author, is given in Chapter 7.

As an integral part of these techniques, we have to determine the order of magnitude of the total horizontal resisting force (the **allowable force** when using the terminology of working-stress design) for different types of structure. In the following section we shall deal with moment-resisting frames.

1.4.2 REINFORCED CONCRETE FRAMES

Let us consider a reinforced concrete moment-resisting frame. The columns are rectangular with cross-sections $b_c \cdot h_c$ (h_c is parallel to the given forces) (Figure 1.31). We assume that:

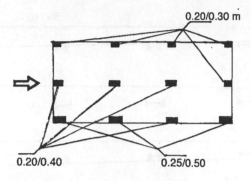

Figure 1.31

- the characteristic (cube) strength of the concrete, f_{ck} varies between 15 and 25 MPa;
- the characteristic strength of the steel, f_y is 300–400 MPa;
- the reinforcement ratio is $\rho = 0.01$;
- the axial (compressive) stresses in the columns vary between 3 and 8 MPa.

By using the graphs prepared for CEB/FIP (1982), we obtain the approximate evaluation of the **resisting moment** M_a due to the horizontal forces F_a in the form (Scarlat, 1993).

$$M_a \cong 0.1 f_{ck} \sum (b_c h_c^2) \qquad (1.36)$$

The sum $\sum$ refers to all the columns; h_c is parallel to the horizontal forces. The total seismic moment acting on the given storey can be expressed in the form

$$M = F \varepsilon h \qquad (1.37)$$

where h is the storey height; ε defines the position of the zero moment point.

We admit $\varepsilon \cong 0.7$ for regular beams and $\varepsilon \cong 1$ for slab-beams. By equating (1.36) and (1.37), we obtain the total **resisting force** acting upon the given frame:

$$F_a \cong \frac{0.1 \sum (b_c h_c)^2 f_{ck}}{\varepsilon h} \qquad (1.38)$$

☐ **Numerical example 1.7**

Determine the total resisting force (allowable force) for the reinforced concrete columns of the moment-resisting frame shown in Figure 1.31. $f_{ck} = 20\,\text{MPa} = 20\,000\,\text{kN}\,\text{m}^{-2}$; $h = 3.20\,\text{m}$; regular beams.

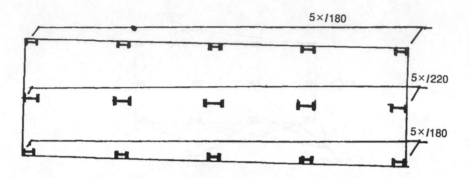

Figure 1.32

$$\sum(b_c h_c^2) = 5 \times 0 \cdot 20 \times 0 \cdot 30^2 + 3 \times 0 \cdot 25 \times 0 \cdot 50^2 + 4 \times 0 \cdot 20 \times 0 \cdot 40^2$$

$$= 0 \cdot 406 \, \text{m}^3; \qquad \varepsilon = 0 \cdot 7$$

$$F_a = \frac{0 \cdot 10 \times 0 \cdot 406 \times 20\,000}{0 \cdot 7 \times 3 \cdot 20} = 362 \, \text{kN} \qquad \qquad \Box$$

1.4.3 STEEL FRAMES

We assume that:

- the steel profiles are St37 (yield strength: 225–235 MPa);
- the assumed resisting stress, $\sigma_a = 220$ MPa $= 220\,000 \, \text{kN m}^{-2}$;
- the assumed average axial (compressive) stress due to vertical forces is $\sigma^N = N/A = 70\text{–}100$ MPa $= 70\,000\text{–}100\,000 \, \text{kN m}^{-2}$.

This yields the resisting stress due to bending moments:

$$\sigma_a^M = 125\text{–}150 \, \text{MPa}; \qquad \text{we admit } \sigma_a^M \cong 130 \, \text{MPa}$$

By assuming $M_a \cong 0 \cdot 7 F h$ (see section 1.4.2), the total resisting horizontal seismic force will be

$$F_a = \frac{M_a}{0 \cdot 7 h} = \frac{\sigma_a^M \sum W_x}{0 \cdot 7 h}$$

$$F_a = \frac{1 \cdot 4 \sigma_a^M \sum W_x}{h} \qquad (1.39)$$

F_a results are in kN when units of kN and m are used and in N when units of N and mm (MPa) are used.

$\sum W_x$ is the sum of the moduli of resistance of the steel profiles and $\sigma_a^M = 130$ MPa $= 130\,000 \, \text{kN m}^{-2}$.

☐ **Numerical example 1.8**

Determine the total resisting horizontal force for the steel moment-resisting frame shown in Figure 1.32.

$$\sum W_x = 5 \times 19 \cdot 8 + 3 \times 278 + 5 \times 19 \cdot 8 = 1588\,\text{cm}^3 \cong 0 \cdot 0016\,\text{m}^3$$

$$\sigma_a^M = 130\,\text{MPa} = 130\,000\,\text{kNm}^{-2}.$$

$$F_a = \frac{1 \cdot 4 \times 130\,000 \times 0 \cdot 0016}{3 \cdot 20} \cong 91\,\text{kN}$$

Bibliography

Allen, F. and Darvall, P. (1977) Lateral load equivalent frame. *Journal of the American Concrete Institute*, **74**, 294–299.

Beck, H. (1956) Ein neues Berechnungsverfahren fur gegliederte Scheiben dargestellt am Beispiel des Vierendeeltragers. *Der Bauingenieur*, **31**, 436–443.

Bleich, F. and Melan, E. (1927) *Die gewohnlichen und partiellen Differenzen Gleichungen der Baustatik*, Springer, Berlin.

CEB/FIP (1982) *Manual on Bending and Compression*, Construction Press, London.

Csonka, P. (1962a) Beitrag zur Berechnung Waagerecht belasteter Stockrahmen. *Bautechnik*, **7**, 237–240.

Csonka, P. (1962b) Die Windberechnung von Rahmentragwerken mit hilfe von Differenzengleichungen. *Bautechnik*, **10**, 349–352.

Ehlers, G. (1957) Die Berechnung von Stockwerkrahmen fur Windlast. *Beton- und Stahlbetonbau*, **52**(1), 11–16.

Klouček, C. (1950) *Distribution of Deformations*, Orbis, Prague.

Kornouhov, N. (1949) *Prochnosti i Ustoichivosti Sterjnevih System*, Stroiizdat, Moscow.

Lightfoot, E. (1956) The analysis for wind loading of rigid-jointed multi-storey building frames. *Civil Engineering*, **51**(601), 757–759; **51**(602), 887–889.

Melnikov, N. and Braude, Z. (1952) *Opit Proektirovania Stalnih Karkasov visotnih zdani*, GIPA, Moscow.

Murashev, V. (1971) *Design of Reinforced Concrete Structures*, MIR, Moscow.

Muto, K. (1974) *Aseismic Design Analysis of Buildings*, Maruzen, Tokyo.

Rutenberg, A. (1966) Multistorey frames and the interaction of rigid elements under horizontal loads. Master Thesis, Technion, Haifa (in Hebrew).

Scarlat, A. (1982) Earthquake resistant design. Lectures, Israel Association of Engineers and Architects (in Hebrew).

Scarlat, A. (1986) Approximate analysis of multistorey buildings in seismic zones. *Premier colloque national de génie parasismique*, St Remy, pp. 6.1–6.10.

Scarlat, A. (1993) First screening of aseismic resistance of existing buildings in Israel. *Handassa Ezrahit iu Binian*, no. 2, 8–15 (in Hebrew).

Williams, F. and Howson, W. (1977) Accuracy of critical loads obtained using substitute frames, in *Proc. EECS, Stability of steel structures*, Liège, April, pp. 511–515.

2 Structural walls

2.1 Introduction

Structural walls represent the most efficient structural element to take lateral forces acting on a multi-storey building and to transfer them to the foundations.

This is borne out by experience of recent earthquakes, which has clearly shown that structures relying on structural walls have been much more successful in resisting seismic forces than structures relying on moment-resisting frames.

Aoyama (1981) notes with reference to the effects of recent earthquakes in Japan that

... the amount of shear walls used played the most important role. This was also the case in previous earthquakes... As to frame buildings or frame portions of buildings, shear failure was the most prevalent type of serious damage or even failure.

Wood (1991), referring to the 1985 earthquake in Viña de Mar, Chile (magnitude 7.8), notes that '... most of these buildings [relying on structural walls] sustained no structural damage.' Fintel (1991) has summarized the effects of 12 strong recent earthquakes and stated that: 'The author is not aware of a single concrete building containing shear walls that has collapsed.'

Structural walls may be either isolated (Figure 2.1a) or form a component of a core (Figure 2.1b). They are either without openings (uniform) or with openings (coupled structural walls).

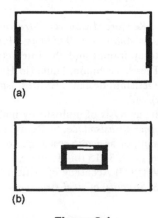

(a)

(b)

Figure 2.1

The main feature of structural walls, as compared with columns of moment-resisting frames, relates to the fact that their rigidities are much more affected by shear forces and soil deformability than are the rigidities of the columns of moment-resisting frames. Both effects will be studied in some detail in the following.

2.2 Structural walls without openings

2.2.1 EFFECT OF SHEAR FORCES

Structural walls without openings may be treated as cantilever beams, but, in contrast with usual cantilevers, where the effect of the bending moments on the deformations is overwhelming (and subsequently the effect of the shear forces may be neglected), the deformations of the structural walls may be strongly influenced by the shear forces; consequently, we have to consider this effect on the deflections, as well as on the rigidities.

In order to quantify this effect, we shall refer to the structural wall shown in Figure 2.2, and compute successively the maximum deflections u_M (by considering only the effect of bending moments) and $u = u_M + u_V$ (by also adding the effect of the shear forces):

$$\left. \begin{aligned} u_M &= \int \frac{m\,M\,\mathrm{d}x}{EI} \\ u = u_M + u_V &= \int \frac{m\,M\,\mathrm{d}x}{EI} + \int \frac{f v\,V\,\mathrm{d}x}{GA} \end{aligned} \right\} \tag{2.1}$$

which give

$$\left. \begin{aligned} u &= u_M \left(1 + \frac{2s}{3}\right) \\ s &= \frac{6fEI}{GH^2A} \end{aligned} \right\} \tag{2.2}$$

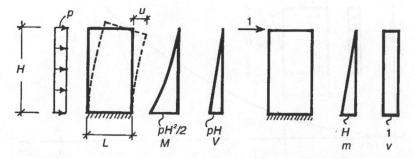

Figure 2.2

f is the **shape factor** depending on the form of the cross-section ($f = 1 \cdot 2$ for rectangles and $2.1-2.5$ for I and $\sqsubset$ sections).

In the case of reinforced concrete rectangular sections and admitting $G = 0 \cdot 425\ E$, we obtain

$$s = 1 \cdot 41\ l^2/H^2; \qquad u = u_M\left(1 + \frac{0 \cdot 94\ l^2}{H^2}\right) \qquad (2.2a)$$

The shear forces lead to a decrease in the rigidity of the structural wall.

The graph shown in Figure 2.3 displays the variation of the ratio u/u_M versus l/H for a rectangular cross-section. We note that for $l/H < \frac{1}{5}$ the effect of the shear forces may be neglected (it represents less than 4%); for $l/H > \frac{1}{3}$, this effect is important (more than 10%); while for $l/H > 1$ it becomes predominant.

Table 2.1 gives:

- the ratios of maximum deflections u/u_M;
- the ratios of fixed end moments $\mathcal{M}/\mathcal{M}_M$;
- the ratios of rigidities K/K_M and carry-over factors t/t_M.

The index (M) signifies that only the effect of bending moments has been taken into consideration. Note that the effect of shear forces appreciably modifies the carry-over factor (Figure 2.4).

In Figure 2.5, we can follow the change of shape of the elastic line of a cantilever subjected to a uniformly distributed load, due to the effect of shear forces.

Note: If we intend to assess the global effect of the shear forces on the stresses and the deflections of a multi-storey building with structural walls, we have to refer to the ratio l/H, where H is the *total* height of the building, and not to the storey height.

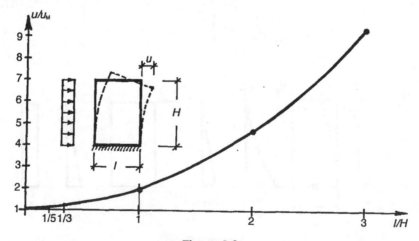

Figure 2.3

Table 2.1 Effect of shear forces on deflections, fixed end moments and rigidities

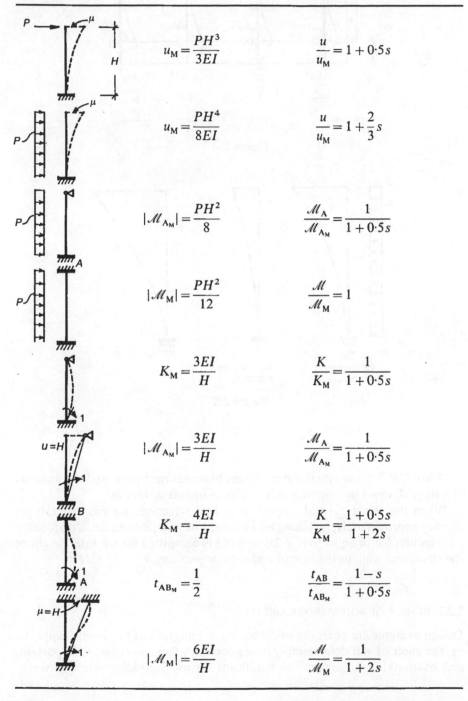

$$u_M = \frac{PH^3}{3EI}$$

$$\frac{u}{u_M} = 1 + 0{\cdot}5s$$

$$u_M = \frac{PH^4}{8EI}$$

$$\frac{u}{u_M} = 1 + \frac{2}{3}s$$

$$|\mathscr{M}_{A_M}| = \frac{PH^2}{8}$$

$$\frac{\mathscr{M}_A}{\mathscr{M}_{A_M}} = \frac{1}{1 + 0{\cdot}5s}$$

$$|\mathscr{M}_M| = \frac{PH^2}{12}$$

$$\frac{\mathscr{M}}{\mathscr{M}_M} = 1$$

$$K_M = \frac{3EI}{H}$$

$$\frac{K}{K_M} = \frac{1}{1 + 0{\cdot}5s}$$

$$|\mathscr{M}_{A_M}| = \frac{3EI}{H}$$

$$\frac{\mathscr{M}_A}{\mathscr{M}_{A_M}} = \frac{1}{1 + 0{\cdot}5s}$$

$$K_M = \frac{4EI}{H}$$

$$\frac{K}{K_M} = \frac{1 + 0{\cdot}5s}{1 + 2s}$$

$$t_{AB_M} = \frac{1}{2}$$

$$\frac{t_{AB}}{t_{AB_M}} = \frac{1 - s}{1 + 0{\cdot}5s}$$

$$|\mathscr{M}_M| = \frac{6EI}{H}$$

$$\frac{\mathscr{M}}{\mathscr{M}_M} = \frac{1}{1 + 2s}$$

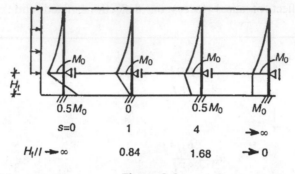

Figure 2.4

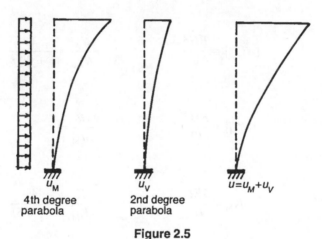

Figure 2.5

When $l/H < \frac{1}{5}$, the effect of shear forces becomes negligible, and the analysis can be performed by referring only to the moments of inertia.

When the structural wall is part of a dual structure, we can evaluate its rigidity approximately, by taking into account its moment of inertia divided by a correction factor equal to $(1 + 2s)$, where s is computed for the total height of the structural wall: in the case of rectangular sections, $s = 1 \cdot 41 \, l^2/H^2$.

2.2.2 EFFECT OF SOIL DEFORMABILITY

Design analyses are generally based on the assumption of fixed ends supports: i.e. the effect of soil deformability is neglected. In fact, this effect is important, and overlooking it may lead to significant errors. Soil deformation strongly affects the rigidity of structural walls and, as such, the magnitude of seismic forces and their distribution between structural walls and frames, as well as the magnitude and distribution of stresses due to change of temperature.

Detailed methods to quantify the soil deformability are given in Appendix C. It was found advisable to use elastic models: a set of discrete springs for large foundations and a 'global' central spring for small foundations.

The basic parameter in defining the soil deformability is the **subgrade modulus** k_s; it usually varies between $20\,000$ kN m^{-3} (soft soils) and $100\,000$ kN m^{-3} (hard soils).

In the following, we shall deal with the effect of soil deformability on the rigidity of structural walls. We shall check the rigidity of the structural walls without openings, by referring to Figure 2.6a.

A **fixed supports analysis** yields the maximum deflection u° and the corresponding rigidity (Figure 2.6b):

$$u = u^\circ; \qquad K_{sw}^\circ = \frac{1}{u^\circ} \tag{2.3}$$

Note: The rigidity of a structure is usually defined as $1/u$, where u is the maximum deflection due to a given lateral load. Here, we considered a concentrated lateral force acting at the top.

An **elastic supports analysis** includes three components of the total displacement u^T (Figure 2.6c, d):

$$u^T = u^\circ + u^H + u^\varphi; \qquad K_{sw} = \frac{1}{u^T} \tag{2.4}$$

where u^H is the supplementary maximum deflection due to horizontal soil deformation and u^φ is the supplementary maximum deflection due to the foundation's rotation (usually $u^\varphi \gg u^H$).

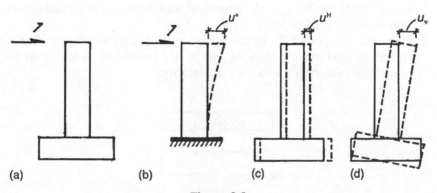

(a)   (b)   (c)   (d)

Figure 2.6

Computations of 0·20 m thick structural walls with heights of 20–30 m on spread footings of 1·50/6·00–2·50/10·00 m have yielded the following results:

$$\text{for soft soils } (k_s = 20\,000\text{–}30\,000 \text{ kN m}^{-3}): \frac{K_{sw}}{K^{\circ}_{sw}} = 0\cdot05\text{–}0\cdot10$$

$$\text{for hard soils } (k_s = 100\,000 \text{ kN m}^{-3}): \frac{K_{sw}}{K^{\circ}_{sw}} = 0\cdot20\text{–}0\cdot30$$

If we replace the elastic supports analysis by a fixed supports analysis, considering structural walls with an **equivalent length** l_{eq} (Figure 2.7), we can obtain the same results:

$$\text{for soft soils: } \frac{l_{eq}}{l} = 0\cdot4\text{–}0\cdot6$$

$$\text{for hard soils: } \frac{l_{eq}}{l} = 0\cdot5\text{–}0\cdot7$$

The decrease in the structural wall rigidities involves an increase in the fundamental period T and a corresponding decrease in the seismic forces acting upon the structural walls. Accepting the provisions of the SEAOC 88 code, which gives a decrease proportional to $1/T^{2/3}$, a decrease in the seismic forces acting on the structure of 35% (hard soils) and 65% (soft soils) is obtained when the structural walls are the only resisting elements. For usual, dual structures (structural walls plus moment-resisting frames), the extent of the decrease is less, according to the rigidity of each type of structural element (see Chapter 3).

From the above-mentioned results it follows that neglecting soil deformability leads to significant errors in the evaluation of rigidities of the structural walls, unacceptable in design. An approximate evaluation of the fundamental period of a structural wall on deformable soil may be obtained as follows.

1. Compute the period T_0 of the structural wall, based on the usual assumption of elastic wall on rigid soil.
2. Compute the period T_r by assuming a rigid wall on deformable soil (subgrade modulus k_s); 'rocking vibrations' (rotations with respect to a horizontal axis) only are taken into account.

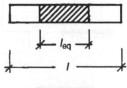

Figure 2.7

3. Evaluate the period T of the elastic wall on deformable soil using the formula [due to Dunkerley (1895)]:

$$T \cong \sqrt{(T_0^2 + T_r^2)} \qquad (2.5)$$

The moment equation for a rigid body in plane motion is:

$$M_0 = I_0 \ddot{\varphi} \qquad (2.6)$$

M_0 denotes the moments of forces about the axis y, and I_0 denotes the polar moment of inertia about the same axis.

For the sake of simplicity, we assume that the structure is symmetrical about the plane yz (Figure 2.8) and uniform. After omitting negligible terms, equation (2.6) becomes

$$\ddot{\varphi} + \omega^2 \varphi = 0$$

where

$$\omega^2 = \frac{g K_\varphi}{W H_T^2/3}$$

yielding

$$T_r = \frac{2\pi}{\omega} = 3 \cdot 63 \sqrt{\left(\frac{W H_T^2}{g K_\varphi}\right)} \qquad (2.7)$$

where W is the total weight of the area tributary to the given structural wall (kN); $K_\varphi = k_s I_f = k_s t_f l_f^3/12$ (kN m rad^{-1}); k_s is the subgrade modulus of the soil (kN m^{-3}), and g is gravitational acceleration (m s^{-2}).

A similar expression is recommended in FEMA 95 (1988). We note that the same expression can also be used in the case of asymmetrical and non-uniform

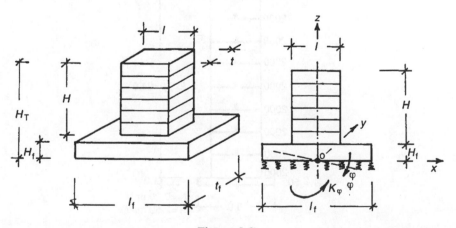

Figure 2.8

structures, without significant errors. In the case of uniform or nearly uniform structural walls with a uniform or nearly uniform distribution of the weights along the height (see section 6.1.4):

$$T_0 \cong 1 \cdot 787 \sqrt{\left(\frac{WH^3}{gEI}\right)}$$

☐ Numerical example 2.1

Consider the eight storey structural wall shown in Figure 2.9. The sum of the weights of the tributary areas is $W = \sum W_i \cong 17\,000$ kN.

(a) Soft soil: $k_s = 20\,000$ kN m^{-3}

The structural wall is uniform. We assume that the weights W_i are uniformly distributed along the height:

$$T_0 \cong 1 \cdot 787 \sqrt{\left(\frac{WH^3}{gEI}\right)}$$

As the cross-section of the foundation is much bigger than the cross-section of the structural wall, we let $H \cong 24$ m.

$$EI = \frac{3 \times 10^7 \times 0 \cdot 2 \times 5 \cdot 0^3}{12} = 62 \cdot 5 \times 10^6 \text{ kN m}^2$$

$$g = 9 \cdot 81 \text{ ms}^{-2}$$

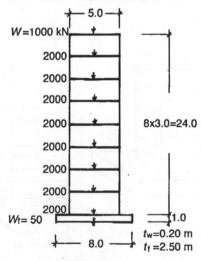

Figure 2.9

$$T_0 \cong 1\cdot787 \sqrt{\left(\frac{17\,000 \times 24^3}{9\cdot81 \times 62\cdot5 \times 10^6}\right)} = 1\cdot11 \text{ s (accurate: } 1\cdot18 \text{ s)}$$

$$T_r \cong 3\cdot63 \sqrt{\left(\frac{WH_T^2}{g\,K_\varphi}\right)}$$

$$K_\varphi = k_s\,I_f = \frac{20\,000 \times 2\cdot5 \times 8\cdot0^3}{12} = 2\,133\,333 \text{ kN m rad}^{-1}$$

$$T_r \cong 3\cdot63 \sqrt{\left(\frac{17\,000 \times 25^2}{9\cdot81 \times 2\cdot133\,333}\right)} = 2\cdot59 \text{ s (accurate: } 2\cdot6 \text{ s)}$$

$$T \cong \sqrt{(T_0^2 + T_r^2)} = \sqrt{(1\cdot11^2 + 2\cdot59^2)} = 2\cdot82 \text{ s (accurate: } 2\cdot89 \text{ s).}$$

(b) *Hard soil* ($k_s = 100\,000$ kN m^{-3}).

$$T_0 \cong 1\cdot11 \text{ s}$$

$$K_\varphi = k_2\,I_f = \frac{100\,000 \times 2\cdot5 \times 8\cdot0^3}{12} = 10\,666\,666 \text{ kN m rad}^{-1}$$

$$T_r \cong 3\cdot63 \sqrt{\left(\frac{17\,000 \times 25^2}{9\cdot81 \times 10\,666\,666}\right)} = 1\cdot16\text{ s (accurate: } 1\cdot17 \text{ s).}$$

$$T \cong \sqrt{(T_0^2 + T_r^2)} = \sqrt{(1\cdot11^2 + 1\cdot16^2)} = 1\cdot61 \text{ s (accurate: } 1\cdot71 \text{ s).}$$

The 'accurate results' were obtained by finite element technique, by considering the structural wall supported on springs. ☐

2.3 Structural walls with openings (coupled walls)

Most structural walls or cores have openings for windows and doors (Figure 2.10). This is an intermediate type of structure, one that is between a

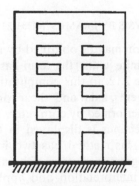

Figure 2.10

structural wall without openings and a moment-resisting frame. The effect of the shear deformations is still important, but it is less than the effect observed in the case of structural walls without openings; it is especially significant in the deformation of lintels, which are usually short and deep. The lintels are the sensitive elements of the structure; they are generally the first to crack during an earthquake, and their proper design is therefore essential for the behaviour of the whole structure.

Moreover, their role in dissipating energy is very important, even after cracking, as they ensure a high ductility of the structure (Paulay and Taylor, 1981). The commentaries to the New Zealand code (1976) point out that

well-proportioned ductile coupled cantilevered shear walls could well be the best earth-quake-resisting structural systems available in RC. The overall behaviour is similar to that of a moment-resisting frame but with the advantage that, because of its stiffness, the system affords a high degree of protection against non-structural damage, even after yielding in the coupling beams.

2.3.1 SLAB COUPLING: EFFECTIVE WIDTH OF LINTELS

Structural walls are in some cases coupled by **slab coupling** (the beams are included in the slab's thickness) (Figure 2.11). In such cases, we have to assess the **effective width** of the lintels to be taken into account in the structural analysis.

According to our computations performed in the elastic range, an effective zone with a width $b_{eff} = 12 t_s + t_w$ (where t_s denotes the slab's thickness and t_w the wall's thickness) can be admitted for the usual range of lintels net spans (1·00−2·50 m), from the points of view of both the deformations and the magnitude of vertical shear forces developed in the lintels.

2.3.2 APPROXIMATE ANALYSIS BY FRAMES WITH FINITE JOINTS

A satisfactory approximation may be obtained by replacing the coupled structural wall by **frames with finite joints** [the term has been coined by I. McLeod (1971)], as shown in Figure 2.12. In this computation we have to consider the effects of the bending moments, axial and shear forces.

The analysis is usually performed by computer. The basic stiffness matrix of members with offset connections is displayed in the technical literature; see Weaver and Gere (1980). This method has two important advantages: it requires a rather simple program (a system of bars instead of finite elements), and allows much simpler interpretation and application of the results than finite elements (we obtain directly the bending moments, the axial and the shear forces acting on the members that we have to proportion).

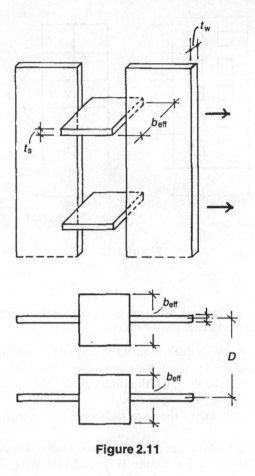

Figure 2.11

□ Numerical example 2.2

The coupled structural wall shown in Figure 2.13a has been replaced by the finite joint frame displayed in Figure 2.13b.

The ratios of the approximate vertical shear forces (V_L^{app}) to the accurate forces (V_L^{acc}), V_L^{app}/V_L^{acc}, are also displayed in Figure 2.13. At most levels $V_L^{app}/V_L^{acc} \cong 0.9$. The maximum displacements are nearly equal:

$$u_{max}^{app} \cong u_{max}^{acc}$$

Considering different component walls, we obtain

$$\text{for} \quad I_2 = 2I_1: \frac{V_L^{app}}{V_L^{acc}} = 0.85\text{--}0.90; \qquad \frac{u_{max}^{app}}{u_{max}^{acc}} \cong 1$$

$$\text{for} \quad I_2 = 8I_1: \frac{V_L^{app}}{V_L^{acc}} = 0.80\text{--}0.95; \qquad \frac{u_{max}^{app}}{u_{max}^{acc}} \cong 0.9$$

In all cases, $x_0/b \cong 0.5$ (zero moment point in lintels).   □

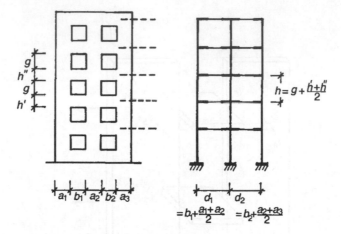

Figure 2.12

2.3.3 APPROXIMATE ANALYSIS OF UNIFORM COUPLED STRUCTURAL WALLS BY THE CONTINUUM APPROACH

Several alternative methods dealing with the analysis of uniform coupled structural walls were developed in the 1960s (Beck, 1962; Arcan, 1964; Rosman, 1966; Penelis, 1969). In the following, we shall refer to the Rosman method, the one with which designers are best acquainted.

The approach is similar to the approach used for uniform one-bay frames (section 1.2.3).

Let us consider a uniform coupled structural wall (all the floors are identical). We replace the actual structure (Figure 2.14a) by a **finite joint frame** (Figure 2.14b). We then replace the lintels, positioned at distances h, by a continuous medium formed by an infinite number of very thin horizontal laminae at distances dx, with the moment of inertia (Figure 2.14c)

$$\frac{I_b\,dx}{h}$$

The laminae are acted upon by infinitesimal vertical shear forces $X\,dx$. Their sum between level 0 (top) and x equals the axial force:

$$N_x = \int X\,dx \qquad (2.8)$$

The equation of compatibility at level x (flexibility method) leads to the differential equation

$$N'' - \alpha^2\,N_x = \psi\,M_{p_x} \qquad (2.9)$$

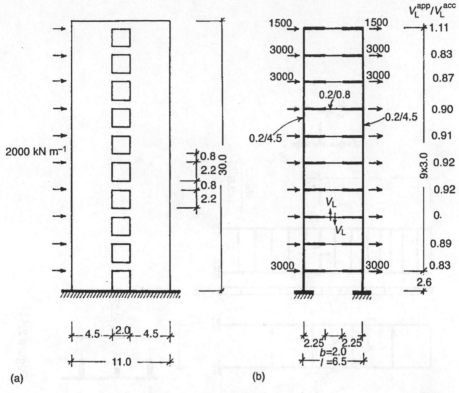

Figure 2.13

where

$$\alpha^2 = \frac{12I_b}{hb^3}\left(\frac{l_0^2}{2I_c} + \frac{2}{A_c}\right) \quad (m^{-2})$$

$$\psi = \frac{6l_0 I_b}{hb^3 I_c} \quad (m^{-3}) \tag{2.10}$$

M_{p_x} is the bending moment acting upon the cantilever wall subjected to the lateral load p_x.

Solving the differential equation (2.9) yields

$$N_X = N_X^{hom} + N_X^{part}; \qquad N_X^{hom} = C \sinh \alpha x + D \cosh \alpha x$$

N_X^{part} depends on the type of loading.

After determining N_X we obtain the shear force X:

$$X = \frac{dN_X}{dx} \tag{2.11}$$

The shear force X_i acting upon the lintel i is shown in Figure 2.14d:

$$X_i \cong Xh \tag{2.12}$$

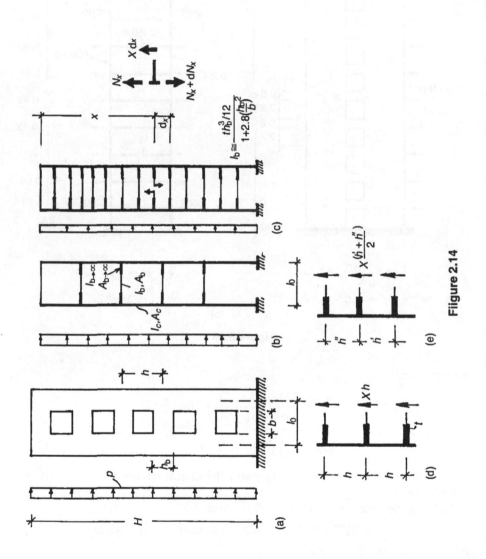

Figure 2.14

When the distances between lintels are variable (Figure 2.14e) we can use the same expression (2.12) by replacing:

$$h = \frac{h' + h''}{2} \tag{2.13}$$

Computation of the shear forces X_i reduces the problem to a statically determinate one, and we can evaluate any stress resultant.

Rosman (1966) has prepared tables for the computation of the shear forces X in the form

$$X = \eta'_x \left(\frac{\psi F_T h}{\alpha^2} \right) \tag{2.14}$$

where F_T is the total lateral force acting upon the structural wall. The coefficient η'_x depends on the type of lateral loading, the coefficient αH and the height x.

The tables prepared by Rosman (1966) include data for uniformly distributed loads, trapezoidal loads and concentrated forces at the top. They have been prepared for $\alpha H = 0.5$–25; when $\alpha H > 25$ we may use $\alpha H = 25$ (η'_x remains practically constant). η'_x is given for the sections $x/H = 0; 0.1; \dots; 0.9; 1$; when the number of storeys $n \neq 10$ we have to draw the diagram V_L through the points $x/H = 0; 0.1; \dots; 1$ and find V_L at the levels of the lintels by interpolation.

☐ **Numerical example 2.3**

The coupled structural wall shown in Figure 2.15a is subjected to a uniform lateral load of intensity

$$p = 20 \, \text{kN m}^{-1} \, (\text{total load: } F_T = 600 \, \text{kN}).$$

$$I_b = \frac{t h_b^3 / 12}{1 + 2.8 h^2 / b^2} = \cdots = \frac{5.893}{10^3} \, \text{m}^4; \qquad l_0 = 6.5 \, \text{m}.$$

The term $2.8 h^2 / b^2$ is intended to take into account the effect of shear forces V_L on the deformations.

$$I_c = \frac{0.2 \times 4.5^3}{12} = 1.519 \, \text{m}^4; \qquad A_c = 0.2 \times 4.5 = 0.9 \, \text{m}^2$$

$$\alpha^2 = 12 \frac{I_b}{h b^3} \left(\frac{l_0^2}{2 I_c} + \frac{2}{A_c} \right) = \cdots = 0.0475 \, \text{m}^{-2}$$

$$\psi = \frac{6 l_0 I_b}{h b^3 I_c} = \cdots = \frac{6.30}{10^3} \, \text{m}^{-3}; \qquad \frac{\psi}{\alpha^2} = 0.133 \, \text{m}^{-1}; \qquad \alpha H \cong 6.5$$

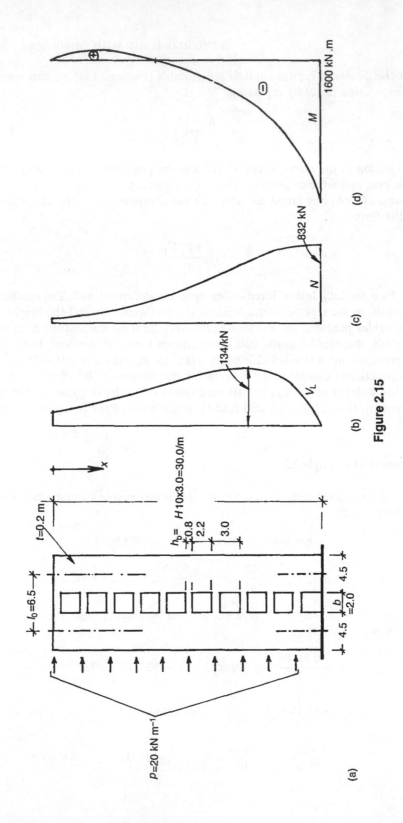

Figure 2.15

The vertical shear forces acting on the lintels are

$$V_{L_x} = \eta'_x \left(\frac{\psi}{\alpha^2} \right) F_T h = \eta'_x \times 0 \cdot 133 \times 3 \times 600 = 239 \cdot 4 \eta'_x$$

η'_x is taken from the tables for uniformly distributed loads and $\alpha H = 6 \cdot 5$. Then we compute

$$N = \sum V_L$$

and

$$M_x = -(p/2) \times (x^2/2) + \frac{N_x l_0}{2} = \cdots = -5x + 3 \cdot 25 N_x$$

The results are shown in Figure 2.15b, c, d. $\square$

2.3.4 ASSESSMENT OF VERTICAL SHEAR FORCES IN LINTELS

In order to assess the order of magnitude of the vertical shear forces in the lintels (V_L) and their distribution we shall refer to three types of lintel:

- flexible lintels – usually a strip of the slab (slab coupling);
- medium lintels – with spans of 2–3 m and heights of 0·50–0·80 m;
- stiff lintels – with spans of 1–2 m and heights of more than 1.00 m (above windows)

Figure 2.16 displays the diagrams of the vertical shear forces V_L developed in the lintels of the coupled structural wall shown in Figure 2.13 along the height, for two types of lateral load: uniformly distributed and inverted triangular.

According to Juravski's formula (by assuming rigid lintels):

$$V_{L,J} = \frac{VSh}{I} = \frac{Vh}{z} \qquad (2.15)$$

where V denotes the horizontal shear force acting upon the cantilever structural wall; S is the first moment of area of the wall with respect to the neutral axis; I is the moment of inertia of the horizontal cross-section with respect to the same axis; and h is the storey height.

$$z = \frac{I}{S} \cong 0 \cdot 7 \, l \quad \text{for structural walls}$$

$$(0 \cdot 8 - 0 \cdot 9) l \quad \text{for cores}$$

l denotes the length of the structural wall/core, parallel to the given forces.

V_L increases with the lintel rigidity; $V_{L,J}$ represents their upper bound. For medium and stiff lintels, the maximum vertical shear forces occur at 0·2–0·3 of the total height measured from the base. For flexible lintels, the vertical shear forces are nearly constant on the upper zone of the shear wall.

The results of several computations are given in the following: the computations were performed for coupled structural walls with 8–12 storeys in order to assess the type of variation and the order of magnitude of the vertical shear

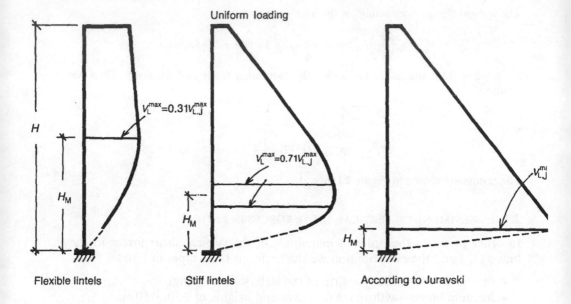

Uniform loading

$V_L^{max}=0.31V_{L,J}^{max}$

$V_L^{max}=0.71V_{L,J}^{max}$

$V_{L,J}^{max}$

Flexible lintels   Stiff lintels   According to Juravski

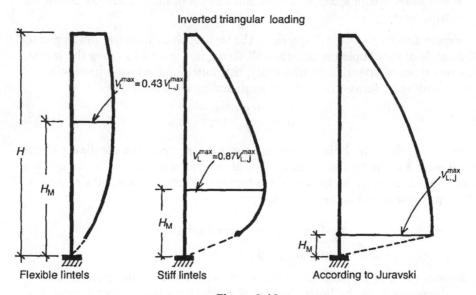

Inverted triangular loading

$V_L^{max}=0.43\,V_{L,J}^{max}$

$V_L^{max}=0.87V_{L,J}^{max}$

$V_{L,J}^{max}$

Flexible lintels   Stiff lintels   According to Juravski

Figure 2.16

forces acting on the lintels. The results are given for fixed ends base (rigid soil). They are compared with the diagram of vertical shear forces $V_{L,J}$ obtained by Juravski's formula. Similar results are given in section 2.3.7 for deformable soils.

(a) Position of the maximum vertical shear force (H_M/H):

Loading type	Flexible lintels	Medium lintels	Stiff lintels
Uniform	0·5–0·7	0·3–0·5	0·2–0·4
Inverted triangular	0·6–0·8	0·4–0·6	0·2–0·4

(b) The sum of the vertical shear forces $\sum V_L$ equals the vertical reaction acting upon each component wall (Figure 2.17a). This can be expressed as a function of the total lateral force F_T. In the following table, the ratios $\sum V_L/F_T$ are given for a coupled structural wall with $H/L = 3$:

Loading type	Flexible lintels	Medium lintels	Stiff lintels
Uniform	0·5–0·9	1·1–1·7	1·5–3
Inverted triangular	0·7–1·5	1·7–2·5	2–4

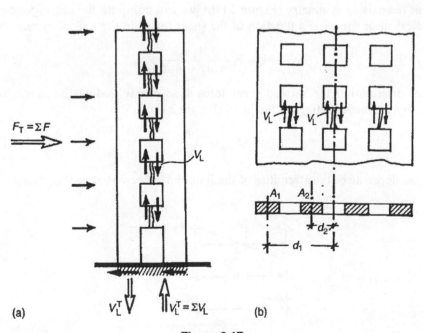

(a)   (b)

Figure 2.17

(c) The intensity of the maximum vertical shear forces V_L^{max} can be expressed as a function of the same shear force computed according to Juravski ($V_L^{max}/V_{L,J}^{max}$):

Loading type	Flexible lintels	Medium lintels	Stiff lintels
Uniform	0·3–0·6	0·6–0·85	0·7–0·9
Inverted triangular	0·2–0·5	0·7–0·85	0·8–0·9

(d) As an alternative, we may assess the same maximum vertical shear force V_L^{max} as a function of the average lateral force $F_{av} = F_T/n$, where n is the number of storeys (V_L^{max}/F_{av}):

Loading type	Flexible lintels	Medium lintels	Stiff lintels
Uniform	0·6–1	1·2–2·5	1·5–3
Inverted triangular	0·8–1·5	1·8–3·5	2–4

The data given above relate to the maximum vertical shear force at the axis of symmetry of each section. When the lintels are positioned at a certain distance from the axis of symmetry (Figure 2.17b), we can compute the corresponding vertical shear forces as a function of the shear forces in the axis of symmetry:

$$V_L' = V_L \frac{A_1 d_1}{A_1 d_1 + A_2 d_2} \tag{2.16}$$

After determining the vertical shear force V_L, we may deduct the maximum bending moments acting on the lintel (Figure 2.18):

$$M_{max} \cong \frac{V_L b}{2} \tag{2.17}$$

(equal slopes at both extremities of the lintel have been assumed: $\varphi_L \cong \varphi_R$).

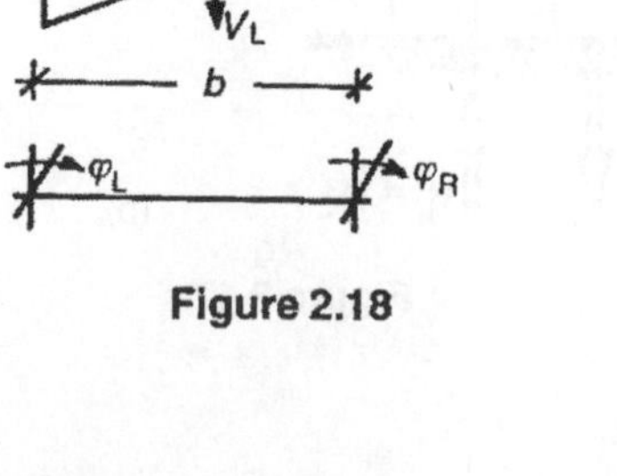

Figure 2.18

Note: Slab coupling is a rather dangerous solution for coupled walls: the early cracking of the lintels significantly decreases the effect of coupling, leading to an important increase in the bending moments acting on each component wall. The results of elastic analyses based on uncracked members are no more valid.

2.3.5 ASSESSMENT OF MAXIMUM HORIZONTAL DEFLECTIONS

If the lintels are considered as infinitely rigid we can determine the maximum deflections (u_{max}°) as for a cantilever with constant moment of inertia (Figure 2.19):

$$I_0 = \frac{I_0' h'}{h} + \frac{I_0'' h''}{h}$$

where

$$I_0' = \frac{tl^3}{12}; \qquad I_0'' = \frac{t(l^3 - b^3)}{12}$$

Owing to the deformability of the lintels:

$$u_{max} > u_{max}^{\circ}$$

The ratio $u_{max} / u_{max}^{\circ}$ depends on the actual rigidity of the lintels, as displayed in the following table (average values for uniformly distributed and inverted triangular loads):

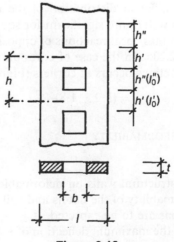

Figure 2.19

Flexible lintels	Medium lintels	Very stiff lintels
3–6	2–4	1.5–2.5

For the numerical example 2.3 (section 2.3.3):

Loading type	Lintels				
	0.5/0.2	1.0/0.2	0.2/0.5	0.2/0.8	0.3/1.0 m
Uniform	5.4	4.2	2.9	2.0	1.7
Inverted triangular	5.5	4.3	2.9	2.0	1.7

2.3.6 DISTRIBUTION OF VERTICAL REACTIONS

Consider a structural wall on rigid supports (Figure 2.20a), subjected to lateral loads.

Usually one assumes a linear distribution of vertical reactions, as a result of the rigid rotation of the structural wall. This leads to the maximum reactions:

$$R_{V_{max}}^0 = \frac{6M}{l} \frac{(n_s - 1)}{n_s(n_s + 1)} \qquad (2.18)$$

where M is the overturning moment (the moment of the resultant lateral loads) at base, and n_s is the number of supports.

In the case of structural walls without openings the distribution of the vertical reactions yielded by an accurate analysis is shown in Figure 2.20b, where $R_{V_{max}} \cong R_{V_{max}}^0$. The different distribution of the reactions is due to the shear deformation of the wall (more significant for squat walls). In the case of structural walls with openings, local reactions of opposite sign may occur close to the openings (Figure 2.20c). In the case of cores on rigid supports a significant concentration of reactions occurs at corners (Figure 2.20d):

$$R_{V_{max}} = (1 \cdot 2 \dots 1 \cdot 8) \, R_{V_{max}}^0 \qquad (2.19)$$

2.3.7 EFFECT OF SOIL DEFORMABILITY

(a) Problems of rigidity

The rigidity of coupled structural walls on deformable soil is affected mainly by two factors: the deformability of the lintels and soil deformability. As such, large horizontal deflections are to be expected.

Let us denote by u_{max} the maximum deflection of a coupled structural wall, and by u'_{max} the corresponding maximum deflection by assuming rigid lintels

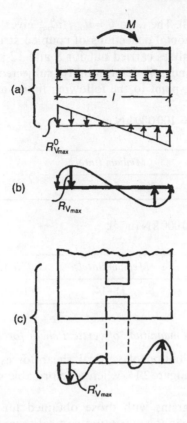

(a)

(b)

(c)

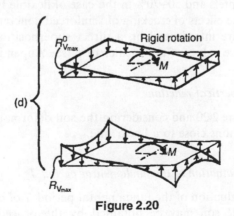

(d)

Figure 2.20

and fixed supports (rigid soil). The ratio $\beta = u_{max}/u^{\circ}_{max}$ gives the order of magnitude of the expected horizontal deflections of coupled structural walls. The results of finite elements analyses carried out for 8 and 12 storeys of coupled walls with various lintels and soil deformability, subjected to uniform and inverted triangular loadings, point to the following limit values of the coefficient β:

Deformable, hard soil ($k_s = 100\,000\,\text{kN m}^{-3}$):

Flexible lintels	Medium lintels	Stiff lintels
6–10	5–9	4–7

Deformable, soft soil ($k_s = 20\,000\,\text{kN m}^{-3}$):

Flexible lintels	Medium lintels	Stiff lintels
12–16	11–14	10–12

(b) Distribution and order of magnitude of vertical shear forces

Figure 2.21 shows the distribution of vertical shear forces for the coupled structural wall displayed in Figure 2.15, when a deformable weak soil is taken into consideration.

By comparing these diagrams with those obtained for fixed ends (Figure 2.15), we note an increase in the maximum vertical shear forces of 10–15% in the case of stiff lintels and 50–70% in the case of flexible lintels. We have to point out that, as the effects of cracking of reinforced concrete elements and of soil deformability are interdependent, a direct superposition of these effects may lead to excessive values of the coefficient β (see Appendix C).

(c) Distribution of vertical reactions

By referring to Figure 2.20 and considering the soil deformability, we obtain a distribution of reactions close to a linear one.

(d) Approximate evaluation of the fundamental period T

An approximate evaluation of the fundamental period T of coupled structural walls on deformable soil may be obtained by the procedure displayed in section 2.2.2, according to the formula

$$T \cong \sqrt{(T_0^2 + T_r^2)} \qquad (2.20)$$

where T_0 is the period computed for the elastic wall on rigid soil and T_r is the period computed for a rigid coupled structural wall on deformable soil.

☐ **Numerical example 2.4**

Let us consider the coupled structural wall shown in Figure 2.22.

$$l = 11 \text{ m}; \quad t = 0 \cdot 2 \text{ m}; \quad H = 30 \text{ m};$$
$$l_f = 14 \text{ m}; \quad t = 2 \text{ m}; \quad H_f = 1 \text{ m};$$
$$E = 3 \times 10^7 \text{ kN m}^{-2} \text{ (Young modulus)}.$$

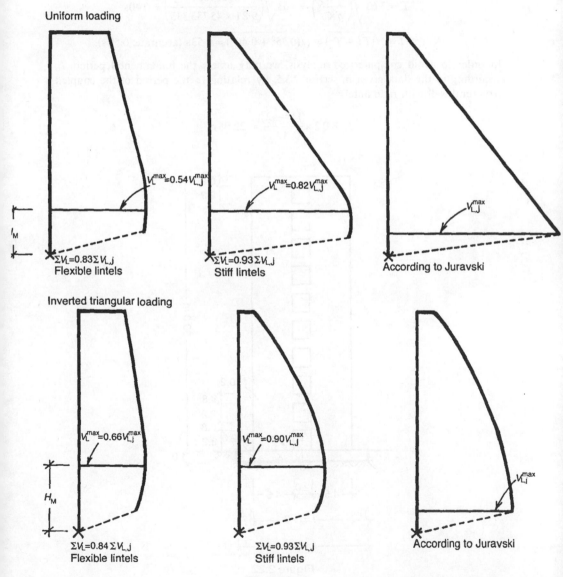

Uniform loading

$V_L^{max} = 0.54 V_{L,j}^{max}$

$\Sigma V_L = 0.83 \Sigma V_{L,j}$
Flexible lintels

$V_L^{max} = 0.82 V_{L,j}^{max}$

$\Sigma V_L = 0.93 \Sigma V_{L,j}$
Stiff lintels

$V_{L,j}^{max}$

According to Juravski

Inverted triangular loading

$V_L^{max} = 0.66 V_{L,j}^{max}$

$\Sigma V_L = 0.84 \Sigma V_{L,j}$
Flexible lintels

$V_L^{max} = 0.90 V_{L,j}^{max}$

$\Sigma V_L = 0.93 \Sigma V_{L,j}$
Stiff lintels

$V_{L,j}^{max}$

According to Juravski

Figure 2.21

Total tributary weight, $W = 5700\,\text{kN}$.

Subgrade moduli to be accounted for: $k_s = 100\,000\,\text{kNm}^{-3}$ and $k_s = 20\,000\,\text{kNm}^{-3}$.
A computerized analysis yields $T_0 = 0\cdot35\,\text{s}$.

(a) $k_s = 100\,000\,\text{kNm}^{-3}$

$$K_\varphi = k_s I_f = \frac{100\,000 \times 2 \times 14^3}{12} = 45\,733\,333\,\text{kNm}\,\text{rad}^{-1}$$

$$T_r = 3\cdot63\,\sqrt{\left(\frac{WH_I^2}{gK_\varphi}\right)} = 3\cdot63\,\sqrt{\left(\frac{5700 \times 31^2}{9\cdot81 \times 45\,733\,333}\right)} = 0\cdot40\,\text{s}$$

$$T = \sqrt{(T_0^2 + T_r^2)} = \sqrt{(0\cdot35^2 + 0\cdot40^2)} = 0\cdot53\,\text{s} \text{ (accurate: } 0\cdot50\,\text{s).}$$

In order to avoid computerized analysis, we may assess the fundamental period T_0 according to the data given in section 2.3.5, by relating to the period of the coupled structural wall with rigid lintels:

$$I_0 = 0\cdot2 \times \frac{11^3 - 2^3}{12} = 22\cdot05\,\text{m}$$

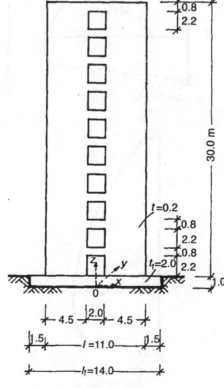

Figure 2.22

For a uniform wall (see Appendix A):

$$T_0 = 1\cdot78 \sqrt{\left(\frac{WH^3}{gEI}\right)} = 1\cdot78 \sqrt{\left(\frac{5700 \times 30^3}{9\cdot81 \times 3 \times 10^7 \times 22\cdot05}\right)} = 0\cdot27\,\text{s}$$

As the lintels are stiff, an increase of $1\cdot3 \ldots 2\cdot5$ in the maximum horizontal deflection is to be expected, yielding an increase of the fundamental period T_0 in proportion to the square root of the deflections: $1\cdot1 \ldots 1\cdot6$; let $T_0 = \frac{1}{2}(0\cdot30 + 0\cdot40) = 0\cdot35\,\text{s}$, resulting in the final period of $T \cong \sqrt{(0\cdot35^2 + 0\cdot40^2)} = 0.53\,\text{s}$ (accurate: $0.50\,\text{s}$).

(b) $k_s = 20\,000\,\text{kNm}^{-3}$

$$K_\varphi = 9\,146\,667\,\text{kNmrad}^{-1}; \qquad T_r = 3\cdot63 \sqrt{\left(\frac{5700 \times 31^2}{9\cdot81 \times 9\,146\,667}\right)} = 0\cdot90\,\text{s}$$

When using the approximate assessment of the period T_0 we obtain:

$$T \cong (0\cdot35^2 + 0\cdot90^2) = 0\cdot97\,\text{s} \quad \text{(accurate: } 0\cdot89\text{s)} \qquad \square$$

2.4 First screening of existing structures laterally supported by RC structural walls

2.4.1 INTRODUCTION

As was shown in section 1.4, we often need a rapid evaluation of the order of magnitude of the seismic capacity of a structure (a detailed review of recent developments in this field is given in Chapter 7).

As a part of this problem we shall now deal with the **first screening** of the seismic capacity of a building laterally supported by reinforced concrete structural walls.

The technique we use is based on concepts set forth by Japanese engineers, derived from statistical studies of the behaviour of numerous buildings during strong earthquakes, especially in the Tokachi area during 1968 and 1969 (Shiga, 1977; Aoyama, 1981). A relationship has been proposed in the form

$$A_{sw} \geqq \frac{0.3}{100} \sum A \tag{2.21}$$

where A_{sw} is the total horizontal area of the reinforced concrete structural walls in a given direction (parallel to the considered seismic forces) at the ground floor level, and $\sum A$ is the sum of the areas of the slabs above the ground floor. Vertical loads of $p = 10\,\text{kNm}^{-2}$ have been assumed. Equation (2.21) was subsequently checked for various earthquakes (Glogau, 1980; Wood, 1991).

Similar relationships are included in the Japanese seismic code (1987) for structures with structural walls and columns. After adjusting the units, they take the form (we refer to ground floor walls):

$$2500\,A_{sw} + 700\,A_c + 1000\,A_{st} > 0\cdot75\,Z_j\,W \quad \text{(kN m)} \tag{2.22}$$

and

$$1800\,A_{sw} + 1800\,A_c > Z_j\,W \quad (kN\,m) \tag{2.23}$$

where A_{sw} was defined above (m^2); A_c is the total area of reinforced concrete columns (m^2); A_{st} is the total area of steel columns (m^2); $W = p\sum A$ is the total weight of the slabs above ground floor; and Z_j is the **hazard seismic coefficient**, admitted in the Japanese code (1987), varying between 0.7 and 1. A rough equivalence with the relative maximum acceleration Z used in the USA may be expressed as $Z_j \cong 2{\cdot}5Z$ (Paulay and Priestley, 1992).

When structures rely on reinforced concrete walls only, equations (2.22) and (2.23) yield

$$3333\,A_{sw} \geqslant Z_j\,W \tag{2.24}$$

$$1800\,A_{sw} \geqslant Z_j\,W \tag{2.25}$$

(units: $kN\,m$).

In this case, only equation (2.25) is significant.

2.4.2 PROPOSED RELATIONSHIPS

Based on the above-mentioned data, the author has proposed the relationship

$$A_{sw} \geqslant \frac{1{\cdot}2}{100}\,Z\,\frac{p}{p_0}\sum A \tag{2.26}$$

where $Z = a_{max}/g$ is the relative peak ground acceleration, given in seismic maps; $p_0 = 10\,kN\,m^{-2}$ is the **reference vertical load**; p $(kN\,m^{-2})$ is the considered vertical load (unfactored). Currently, $p \cong p_0$, leading to

$$A_{sw} \geqslant \frac{1{\cdot}2}{100}\,Z\sum A \tag{2.27}$$

$Z = 0{\cdot}2\ldots0{\cdot}3$ yields

$$A_{sw} \geqslant \frac{0{\cdot}24}{100}\sum A \ldots \frac{0{\cdot}36}{100}\sum A$$

close to the basic relationship (equation (2.21)). It is consistent with data obtained from the 1985 earthquake in Viña del Mar, Chile (Wood, 1991).

We now consider the approximate total shear force derived from the SEA-OC-88 code by considering an average soil, normal importance of the building and referring to unfactored seismic forces:

$$V \cong \frac{3ZW}{\sqrt{H}} = \frac{3Zp\sum A}{\sqrt{H}} \tag{2.28}$$

where H is the total height of the structure (m).

By considering an average storey height of 3 m and n identical storeys:

$$H = 3n; \qquad \sum A = n A$$

yielding

$$V = 1.73 \sqrt{n} Z p A \qquad (2.29)$$

The **resisting (allowable) base shear** V_a results in

$$V_a = A_{sw} \tau_a = \frac{1.2}{100} Z \sum A \tau_a = \frac{1.2}{100} Z n A \tau_a \qquad (2.30)$$

where τ_a is the allowable shear stress of the structural walls.

By equating (2.29) and (2.30) we obtain

$$\tau_a = \frac{144}{\sqrt{n}} p \qquad (2.31)$$

τ_a decreases with the number of storeys.

When a load $p = 10\,\text{kN}\,\text{m}^{-2}$ is assumed:

$$
\begin{aligned}
n = 2 \quad & \tau_a = 1020\,\text{kN}\,\text{m}^{-2} = 1\cdot0\,\text{MPa} \\
4 \quad & = 720 \qquad\qquad\quad = 0\cdot7 \\
6 \quad & = 588 \qquad\qquad\quad = 0\cdot6 \\
8 \quad & = 509 \qquad\qquad\quad = 0\cdot5 \\
10 \quad & = 455 \qquad\qquad\quad = 0\cdot45
\end{aligned}
$$

For $n \geqslant 12$ a constant value $\tau_a = 0.4\,\text{MPa}$ is admitted.

Note: When computing the total horizontal area of reinforced concrete structural walls A_{sw} we take into account walls at least 0·12 m thick and at least 1·00 m long (parallel to the given forces). Walls with less than 1 m length are considered as columns. In the case of coupled walls we count for the whole length L_{sw} (Figure 2.23) when the height of the lintels h_b (including the slab) exceeds 0·70 m; otherwise the length $2a$ is considered.

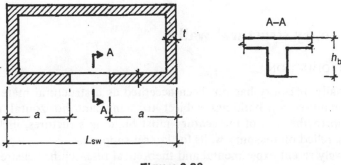

Figure 2.23

In the case of cores, we take into account in each direction the total length of the walls included in the core (parallel, as well as normal, to the given forces).

In order to compare the results of the proposed relationships with the results given by the Japanese code, we consider equation (2.25), which yields

$$1880\, A_{sw} \geqslant Z_j\, W = Z_j\, p \sum A \tag{2.32}$$

As $Z_j \cong 2 \cdot 5\, Z$

$$1880\, A_{sw} > 2 \cdot 5\, Z\, p \sum A \tag{2.33}$$

When $p = 10\,\text{kN}\,\text{m}^{-2}$

$$A_{sw} \geqslant \frac{1 \cdot 3}{100} Z \sum A = \frac{1 \cdot 3}{100} Z \sum A \tag{2.34}$$

(compared with the proposed value $A_{sw} \geqslant (1 \cdot 2/100)\, Z \sum A$).
The allowable shear stress yielded by equation (2.32) is:

$$\tau_a' = 1880\,\text{kN}\,\text{m}^{-2} = 1 \cdot 88\,\text{MPa}$$

for a conventional force $V' = Z_j\, np\, A = 2.5\, Z np\, A$, much higher than the force we considered: $V = 1 \cdot 73 \sqrt{n Z p A}$ (equation (2.22)).
By adjusting τ_a' accordingly:

$$\tau_a = 1 \cdot 88 \frac{1 \cdot 73 \sqrt{n Z p A}}{2 \cdot 5 n Z p A} \cong \frac{1 \cdot 3}{\sqrt{n}} \,\text{MPa}$$

leading to:

$$n = 4: \quad \tau_a = 0 \cdot 65\,\text{MPa} \quad \text{(compared with } 0 \cdot 7\,\text{MPa)}$$

$$n = 10: \quad \tau_a = 0.41\,\text{MPa} \quad \text{(compared with } 0 \cdot 45\,\text{MPa).}$$

We note that the seismic coefficients resulting from the Japanese formulae do not vary with the building's height as in the SEAOC-88 code used in our proposed relationships.

2.5 Masonry structural walls

2.5.1 INTRODUCTION

Traditionally, masonry has not been accepted as a structural material fit for modern multi-storey buildings subject to significant horizontal forces, although up to the turn of the century most building structures, including tall buildings, relied on masonry walls for lateral support.

Relatively recent experimental and theoretical research has cleared the way for considering masonry walls as a recognized material, alongside steel and reinforced concrete.

In the case of usual structures, i.e. in reinforced concrete or steel, the masonry walls, either facades or partitions, are not considered by most designers as structural elements. Neglecting masonry walls has three main effects.

- It lowers the actual rigidity of the structure and subsequently leads to underestimation of seismic forces.
- It significantly modifies the position of the centre of rigidity (see Chapter 4) and subsequently the magnitude of the general moment of torsion.
- It neglects important reserves of resistance. Numerous cases are cited in the technical literature where brick walls acting together with RC elements have saved buildings from collapse during seismic attacks (Anon, 1982; Clough *et al.* 1990; Gulkan *et al.* 1990).

Several authors, as well as codes (e.g. CEB, 1985) recommend separating masonry walls from the main structural elements, in order to achieve a clear statical scheme and to avoid undesirable eccentricity. However, we do not agree with such proposals, as separating these walls leads to a significant decrease in the capacity of the structure and to possible collapse due to out-of-plane forces. We also have to recognize that the means of separation (usually by introduction of deformable materials such as polystyrene) are not sufficient to ensure a clear separation.

In the following, we shall refer to three main types of masonry walls: plain masonry (unreinforced and not connected to a grid of reinforced concrete or steel columns and beams); infilled frames (the infill masonry is not reinforced, but is connected to a reinforced concrete grid of columns and beams); reinforced masonry.

The main data for determining the strength and the deformability of brick masonry can be found in specific regulations, e.g. in UBC-88. They are as follows.

(a) Strength data

The specified compressive strength (f'_m) varies as a function of the quality of the mortar and the level of inspection, between 10 and 40 MPa for clay masonry units, and between 6 and 20 MPa for concrete masonry units. The recommended limits for seismic zones where $Z = 0.3$ are $f'_m = 18$ MPa for clay units and $f'_m = 10.5$ MPa for concrete units.

The allowable compressive stress (flexural) $f_m = 0.33f'_m$.

The allowable shear stress τ_{a_m} (F_v in UBC-88 notation) in unreinforced masonry:

- clay units: $\tau_{a_m} = 0.025 \sqrt{f'_m}$ and $\tau_{a_m} = 0.56$ MPa
- concrete units: $\tau_{a_m} = 0.24$ MPa;
- stone with cement mortar: $\tau_{a,sm} = 0.03-0.06$ MPa, as a function of the level of inspection.

(b) Deformability data

Modulus of elasticity for uncracked masonry, $E_m = 750 f'_m$ for concrete masonry and $E_m = 1000, f'_m$ for clay brick masonry. Shear modulus $G = 0\cdot4 E$.

For cracked masonry, Hart (1989) suggested multiplying the modulus of elasticity by $0\cdot4$.

2.5.2 PLAIN MASONRY (IN-PLANE BENDING)

The use of plain masonry as a structural element in seismic areas is not recommended. Consequently, we have to deal only with existing plain masonry structures, in order to evaluate their seismic resistance.

We have to point out the uncertainty stemming from the questionable quality of existing plain masonry structures (chiefly the mortar strength), and from the conflicting data cited in the technical literature dealing with the allowable stresses of plain masonry walls. It is therefore advisable to adopt very approximate – and usually very conservative – methods of analysis.

By referring to data used for 'first screening' techniques (see section 2.7 and Chapter 7) we recommend **resisting (allowable) base shear** $V_{a,m}$ in a given direction in the form

$$V_{a,m} = \tau_{am} A_m$$

where A_m is the total horizontal area of the structural masonry walls in the given direction. Several limitations in defining the area A_m are detailed in section 2.7.1.

The allowable shear stress τ_{am} can be taken between $0\cdot03$ and $0\cdot05$ MPa for solid unit masonry, and between $0\cdot015$ and $0\cdot03$ MPa for hollow unit masonry, as a function of the present condition of the masonry.

2.5.3 INFILLED FRAMES (IN-PLANE BENDING)

Several techniques have been proposed to evaluate the allowable horizontal force of an infilled frames masonry wall subject to in-plane bending and axial force. We have chosen the procedure proposed by Stafford Smith and Coull (1991).

Three failure modes are studied:

- tension failure mode, involving yield of the steel in the tension column;
- compression failure mode, involving failure dictated by the compression strength of the diagonal strut of the panel (Figure 2.24);
- shear failure of the masonry infill, leading to the onset of flexural plastic hinges in the RC columns (Figure 2.25).

As the second and the third failure modes are usually more dangerous, we shall deal only with these modes.

The resisting (allowable) base shear for compression failure may be expressed in the form

$$V_{a,m} = 2 f_m \sum \left[\cos^2 \theta (I_c h_m t^3)^{1/4} \right] \ (\text{MPa, mm}) \tag{2.35}$$

for RC frames, and

$$V_{a,m} = 3f_m \sum [\cos^2\theta(I_c h_m t^3)^{1/4}] \quad (\text{MPa, mm}) \tag{2.36}$$

for steel frames; $\tan\theta = h_m/L_m$.

f_m is the allowable compressive stress (MPa). The sum $\sum$ refers to all the infilled panels included in the considered wall; L_m is the length of the infill; t and h_m denote the thickness and the height of a masonry panel; I_c denotes the moment of inertia of each column bounding a masonry panel. The sum $\sum$ extends to all the panels included in the wall.

The **resisting (allowable) base shear** $V_{a,m}$ for shear failure may be expressed in the form

$$V_{a,m} = \sum \left(\frac{L_m t}{1\cdot39 - 0\cdot16\,h_m/L_m}\right)\tau_{am} \quad (\text{MPa, mm}) \tag{2.37}$$

τ_{am} is the allowable shear stress and L_m is the length of the panel. In evaluating $V_{a,m}$, an internal friction coefficient of $0\cdot2$ was assumed.

☐ **Numerical example 2.5**

Check the infilled RC frame wall shown in Figure 2.26, at the ground floor.

$$t = 200\,\text{mm}; \qquad h_m = 2\cdot80\,\text{m}$$

Allowable shear stress $\tau_{am} = 0\cdot24$ MPa

Allowable compressive stress $f_m = 7$ MPa

Base shear $V = 400$ kN

We shall compute the resisting base shear $V_{a,m}$.   ☐

(a) Compressive failure

Zones (1) and (2): $L_m = 4700\,\text{mm}$; $h_m = 2800\,\text{mm}$; $\tan\theta = 2800/4700 = 0\cdot596$; $\cos\theta = 0\cdot859$.

$$I_c = \frac{200 \times 300^3}{12} = 4\cdot5 \times 10^8\,\text{mm}^4$$

Zones (3): $L_m = 3800\,\text{mm}$; $h_m = 2800$; $\tan\theta = 0\cdot737$; $\cos\theta = 0\cdot805$.

$$I_c = \frac{200^4}{12} = 1\cdot33 \times 10^8\,\text{mm}^4$$

$$V_{a,m} = 2 \times 7[2 \times 0\cdot859^2 \times (4.5 \times 10^8 \times 2800 \times 200^3)^{1/4}$$

$$+ 0\cdot805^2 \times (1\cdot33 \times 10^8 \times 2800 \times 200^3)^{1/4}] = 1\,541\,072\,\text{N}$$

$$\cong 1541\,\text{kN}$$

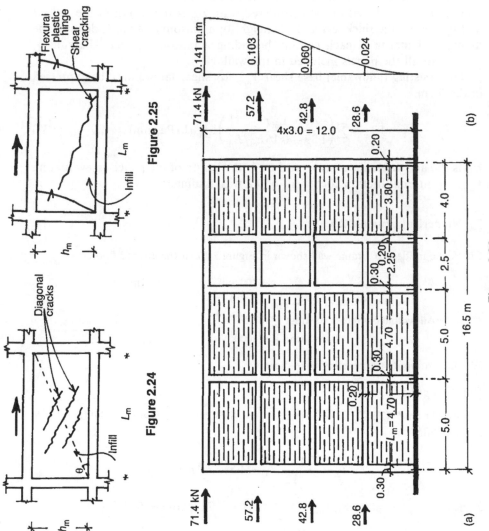

Figure 2.25

Flexural plastic hinge

Shear cracking

Infill

L_m

h_m

Figure 2.24

Diagonal cracks

Infill

L_m

h_m

θ

Figure 2.26

(a)

71.4 kN

57.2

42.8

28.6

$L_m = 4.70$

0.30

0.20

0.30

4.70

0.30 0.20

2.25

3.80

0.20

5.0

5.0

2.5

4.0

16.5 m

(b)

71.4 kN

57.2

42.8

28.6

4x3.0 = 12.0

0.141 m.m

0.103

0.060

0.024

(b) Shear failure

$$V_{a,m} = 0.24 \times \left[2 \times 4700 \times \frac{200}{(1.39 - 0.16 \times 2800/4700)} \right.$$
$$\left. + 3800 \times \frac{200}{(1.39 - 0.16 \times 2800/3800)} \right] = 491\,887\,\text{N} \cong 492\,\text{kN}$$

$$V_{a,m} = 492\,\text{kN} > V = 400\,\text{kN}$$

The structural wall can resist the given lateral forces.   □

Note: Deflections can be computed by replacing the infilled panels by trusses formed by RC members (columns and beams) and masonry diagonal struts (Figure 2.27) with a width $d_m = L_d/4$ (Wakabayashi, 1986). Their moments of inertia are: $t d_m^3/12$; the modulus of elasticity $E_m = 750 f'_m$.

□ **Numerical example 2.6**

By referring to the infilled frame of Figure 2.26 we obtain the truss shown in Figure 2.28, intended to permit the computation of horizontal deflections.

RC beams: 1200/200 mm, columns: 200/300 (left zone), 200/200 (right zone).
Masonry diagonal struts: 1460/200 mm (left zone), 1250/200 (right zone).
RC elements: $E = 30\,000$ MPa; masonry: $E_m = 16\,000$ MPa.

Computer analysis yields the elastic line displayed in Figure 2.26b.   □

As is customary in design analyses we assume elastic, uncracked members for masonry as well as for reinforced concrete, but we must be aware that, in the case of strong earthquakes, the rigidity of infilled frames significantly decreases owing to cracks in the RC elements, as well as in the masonry infill.

In the following, we shall refer to two tests performed in order to determine the effect of the brick infill on the stiffness and the failure load of an infilled frame.

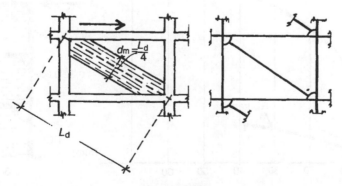

Figure 2.27

Tests with statically applied forces have been performed in Johannesburg on a three-storey RC frame (cited by Stafford Smith and-Carter, 1969) (Figure 2.29), yielding:

$$u/H = 1/1000 \quad F_{\text{infill}}/F_{\text{RC}} = 7\cdot4$$
$$1/500 \qquad\qquad = 5\cdot1$$

$$1/300 \qquad\qquad = 4\cdot7 \text{ (near failure of the infilled frame)}$$

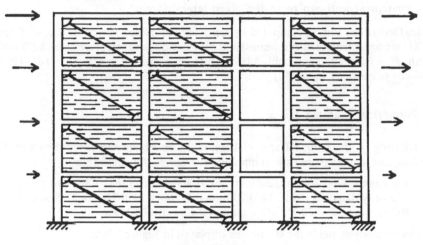

Figure 2.28

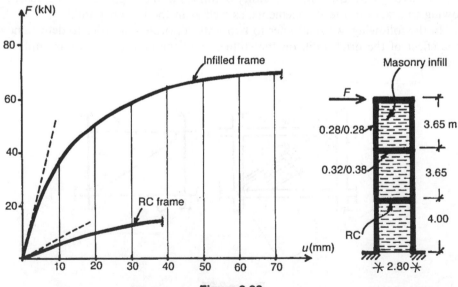

Figure 2.29

where H denotes the total height and u the maximum horizontal deflection. Note that the infilled frame exhibits a remarkably high strength up to near failure.

Cyclic load tests have been performed on a 1/4 scale model of a seven-storey building in two alternatives: a RC frame and an infilled frame (Govindan *et al.*, 1986). The effect of infill on rigidity is significant only during the first earthquake shocks, and subsequently nearly vanishes, in spite of the fact that the failure load of the infill frame remains much higher than the failure load of the RC frame (approximately 2.7 times higher).

This is interpreted as indicating that the parasitic torsional moments due to the collaboration of the masonry wall with the RC structure may be significant only during the first earthquake shocks, when the structural rigidity is high due to the infilled frames. When the first cracks set in, the increase in the rigidity due to the masonry walls is less important (and the parasitic torsional moments decrease as well), but the increase in the strength of the masonry remains important and beneficial to the structure as a whole. This confirms our view that the interaction between the main structure (steel or RC) and masonry walls should be maintained.

2.5.4 REINFORCED MASONRY (IN-PLANE BENDING)

(a) Design for bending moments and axial forces

Proportioning of reinforced masonry is usually performed by an elastic approach. Paulay and Priestley (1992) proposed that the proportioning be based on the flexural strength of the wall, similar to that of proportioning of RC sections. They showed that such an approach is also reasonable for reinforcing bars uniformly distributed along the wall. In the following, we shall adopt this approach; we shall determine the **resisting (allowable) bending moment** $M_{a,m}$ by considering a constant axial force N.

Referring to Figure 2.30 it may be assumed that, at failure, most of the reinforcing bars reaching the yield point (f_y) are in tension, while only a few are in compression.

We denote:

$$C_b = a t f'_m = 0.85 \, c t f'_m \qquad (2.38)$$

(when the masonry is confined by steel plates, we let $a = 0.96 c$ instead of $a = 0.85 c$).

$$C_s = n_c A_s f_y \qquad (2.39)$$

$$T_s = n_t A_s f_y$$

where A_s denotes the area of one reinforcing bar.

The total number of bars $n = n_t$ (in tension) $+ n_c$ (in compression).

Equivalence equations of the resultant stresses $M_{a,m}$ (allowable moment), axial force N and the forces C_b, C_s, T_s yield

$$N = C_b + C_s - T_s$$
$$= 0.85\,c\,t f'_m + n_c\,A_s f_y - n_t\,A_s f_y \qquad (2.40)$$

$$M_{a,m} = C_b(c - 0.5\,a) + A_s f_y \sum (c - d_i) + N(0.5\,L_m - c) \qquad (2.41)$$

Equation (2.40) yields

$$c = \frac{N + (n_t - n_c)\,A_s f_y}{0.85\,t f'_m} \qquad (2.42)$$

An iterative computation is performed by initially choosing a given number of compressed bars (n_c); usually $n_c = 1$ or 2.

☐ **Numerical example 2.7**

See Figure 2.30. Concrete masonry blocks, $t = 200$ mm thick; $L_m = 15 \times 0.20 = 3.0$ m. Specified compressive strength $f'_m = 8$ MPa. $n = 8$ steel bars D16 mm, $f_y = 400$ MPa; for each bar $A_s f_y = 200 \times 400 = 80\,000\,N = 80$ kN. We shall compute the resisting moment $M_{a,m}$.

We assume $n_c = 2$; then $n_t = 8 - 2 = 6$. According to equation (2.42):

$$c = \frac{20\,000 + (6 - 2)\,200 \times 400}{0.85 \times 200 \times 8} = 250 \text{ mm}$$

$$a = 0.85\,c = 213 \text{ mm}$$

We assume that the compressive zone includes only one steel bar: $n_c = 1$; $n_t = 8 - 1 = 7$.

$$c = \frac{20\,000 + (7 - 1)\,200 \times 400}{0.85 \times 200 \times 8} = 368 \text{ mm}$$

$$a = 0.85\,c = 313 \text{ mm}$$

The compressive zone includes one steel bar, as assumed; no additional iteration is needed.

$$C_b = a\,t f'_m = 313 \times 200 \times 8 = 500\,800 \text{ N} \cong 501 \text{ kN}$$

As equations (2.40) and (2.42) are equivalent, equation (2.40) is satisfied. According to equation (2.41) the resisting moment results in

$$M_{a,m} = 501\,(0.37 - 0.5 \times 0.31) + 80 \times [(0.37 - 0.10) + (0.50 - 0.37)$$
$$+ \cdots + (2.90 - 0.37)] + 20\,(0.5 \times 3.00 - 0.37) = 897 \text{ kN m} \qquad ☐$$

(b) Design for shear forces

The total resisting shear force:

$$V_{a,T} = V_{a,m} + V_{a,sh} \qquad (2.43)$$

where $V_{a,m}(V_{a,sh})$ denote the resisting shear forces taken by the masonry (horizontal steel bars).

According to Shing *et al.* (1989), cited by Paulay and Priestley (1992):

$$V_{a,m} = \tau_a A_m \qquad (2.44)$$

where

$$\tau_a \leqslant 0.04 \sqrt{f'_m} + \frac{0.16 \, N}{A_m} \quad \text{(MPa)}$$

$$\tau_a \leqslant 0.20 + \frac{0.16 \, N}{A_m} \quad \text{(MPa)}$$

$$\tau_a \leqslant 0.52 \quad \text{(MPa)}$$

$$A_m = L_m t$$

$$V_{a,s} = 0.8 f_y \cdot A_{sh} \qquad (2.45)$$

where A_{sh} denotes the total area of the horizontal steel bars of the considered wall.

2.6 Out-of-plane bending of masonry walls

In evaluating the out-of-plane seismic forces (normal to the wall plane) we can use the formula given by the SEAOC-88 code in the form

$$F = \frac{ZIC}{R_w} W_m \qquad (2.46)$$

where W_m denotes the weight of the masonry wall and I is the importance factor ($I = 1 \ldots 1.25$).

The maximum forces will develop at the top storeys. We admit, conservatively:

$$C = C_{max} = 2.75$$

$$R_w = R_{w_{min}} = 1$$

resulting in

$$F = 2.75 \, ZI \, W_m \qquad (2.47)$$

The wind pressure on masonry partition walls varies, according to national codes, between 0.3 and 0.5 kN m^{-2}.

When plain masonry walls of existing buildings are checked for out-of-plane forces, and we cannot ascertain the existence of an accepted type of connection to the main structural elements (steel or RC columns), the situation may be very dangerous; such masonry walls sustaining vertical loads must be strengthened. In the case of infilled frames, where a connection exists between the infill and the structural elements, we check the infill acted upon by out-of-plane forces, by taking into account the existing supports.

It should be pointed out that the actual resistance exhibited by the infill masonry connected to structural elements is significantly higher than the resistance resulting from usual design. This can be explained by the **arching effect**: i.e. the masonry behaves like a compressed arch fixed at the extremities (Figure 2.31).

In the case of reinforced masonry, the existing reinforcing bars, even when corresponding to minimum reinforcement ratio, usually suffice to ensure resistance to out-of-plane forces.

2.7 First screening of existing buildings laterally supported by masonry walls

In the case of masonry walls, we shall compute the resisting (allowable) base shear $V_{a,m}$ in the form

$$V_{a,m} = \tau_{am} A_m \qquad (2.48)$$

As usual in 'first screening' procedures, we refer to existing masonry walls at the ground floor only.

$A_m = \sum L_m t$ denotes the total horizontal area of the brick masonry at the ground floor in the given direction. Its computation is subject to several restrictions (Figure 2.32): masonry walls with at least 1 m length and 150 mm thick are taken into consideration; the masonry above the doors is neglected; windows are included in the area A_m. When stone masonry is checked we take into consideration only walls at least 200 mm thick.

Different allowable shear stresses $\tau_{am}(\tau_{a,sm})$ are prescribed for each type of masonry wall (stone masonry wall).

2.7.1 PLAIN MASONRY

As stated earlier (section 2.5.3), the quality of existing plain masonry and, especially, the quality of the mortar are very uncertain, and we must therefore be cautious in assessing its seismic resistance.

The allowable shear stress $\tau_{a,m}$ will be chosen according to data provided by the codes and handbooks. FEMA-178 (1989) recommends values of 0.07 MPa for solid concrete masonry and 0·04 MPa for hollow unit masonry. When vertical compressive stresses of the masonry are neglected, Hendry (1990) recommends $\tau_{u,m} = 0·3$ MPa (ultimate strength); by letting $\tau_{a,m} = 0·1\ \tau_{u,m}$ [according to UCBC (1992)], we obtain $\tau_{a,m} = 0·03$ MPa. UBC-88 recommends

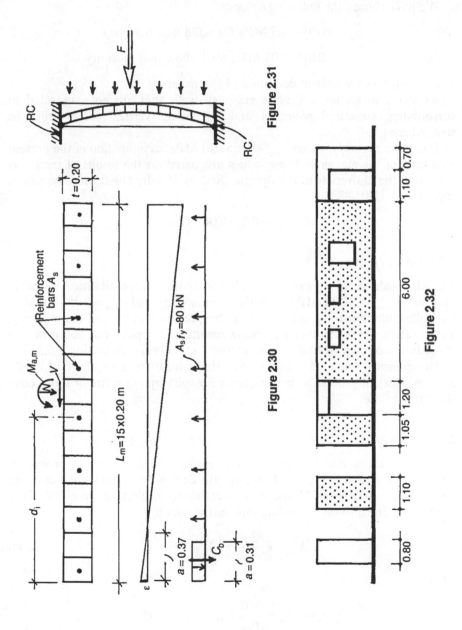

Figure 2.31

Figure 2.30

Figure 2.32

$\tau_{a,m} = 0.04-0.08$ MPa for solid unit masonry and $\tau_{a,m} = 0.035-0.07$ MPa for hollow unit masonry.

We have chosen the following values:

$$\tau_{a,m} = 0.03-0.05 \text{ MPa for solid unit masonry}$$

$$0.015-0.03 \text{ MPa for hollow unit masonry}$$

as a function of the present condition of the masonry.

As stated in section 2.6, plain masonry walls that are not connected to surrounding structural members and sustaining vertical loads must be strengthened.

For stone masonry, we let $\tau_{a,sm} = 0.05-0.1$ MPa, as a function of the present condition of the masonry. These values are based on the results of tests performed at the University of Edinburgh, cited by Hendry (1990), and the values specified by UBC-88:

$$\tau_{a,sm} = 0.03-0.06 \text{ MPa}.$$

2.7.2 INFILLED FRAMES

The allowable shear stresses are much higher than those admitted for plain masonry: $\tau_{a,m} = 0.1-0.2$ MPa for solid unit masonry and $\tau_{a,m} = 0.05-0.1$ MPa for hollow unit masonry, depending on the present condition of the masonry. These values were determined by using results of computations based on the procedure proposed by Stafford Smith and Coull (1991) (see section 2.5.3) and on the empirical rule used in Japan (Wakabayashi, 1986): a length of 0.15-0.20 m of masonry at ground floor is required for each square metre of slab above the ground floor.

2.7.3 REINFORCED MASONRY

In order to assess the allowable shear stress $\tau_{a,m}$ we shall use the procedure described in section 2.5.4, based on strength design, by assuming a minimum steel ratio $\rho = \rho_{min} = 0.07\%$, and by conservatively neglecting the effect of the axial force. The allowable bending moment results in

$$M_a = F_a \frac{2H}{3} = \left(\sum A_s\right) f_y z \qquad (2.49)$$

where

$$\sum A_s = \rho A_m = \frac{0.07}{100} A_m; \qquad z \cong 0.45 L_m$$

L_m denotes the length of the wall, and H is the total height of the masonry wall.

The allowable horizontal force takes the form

$$F_{a,m} = \tau_{a,m} A_m \tag{2.50}$$

The allowable shear stress results in

$$\tau_{a,m} = \frac{5}{10^4} \quad f_y = \frac{L_m}{H} \tag{2.51}$$

2.8 Connection forces between facade panels in one-storey industrial buildings

In the case of one-storey industrial buildings, the horizontal forces are often carried to the foundations by precast facade panels. It is important to assess the order of magnitude of the connection forces that develop between these panels subjected to horizontal loads (Scarlat, 1987).

Let us consider a wall of this type supported by a number of isolated foundations (spread footings or piles); they are replaced by a continuous foundation (Figure 2.33). The top face of the foundation is acted upon by horizontally distributed shear forces $p_H = F/L$ and distributed moments $m = p_H h$. The bottom face is acted upon by the vertical forces $p_v = \sigma b$ (b denotes the width of the foundation beam).

The moment equilibrium yields

$$p_{max} = \sigma_{max} b = \frac{Mb}{bL^2/6} = 6F\frac{h}{L^2}$$

$$p_x = p_{max}\frac{x}{L/2} = 12Fh\frac{x}{L^3}$$

$$P_x = \frac{p_x x}{2} = \frac{6Fhx^2}{L^3}$$

$$V_x = V_s - P_x = 1{\cdot}5\frac{Fh}{L} - 6\frac{Fhx^2}{L^3}$$

$$M_x = P_x\frac{x}{3} + mx - V_s x = Fh\frac{x}{L}\left(2\frac{x^2}{L^2} - 0{\cdot}5\right)$$

$$\left. \begin{array}{l} M_{max} = 0{\cdot}1Fh \ (\text{at } \bar{x} = 0{\cdot}29\,L) \\[2ex] V_{corresp} \cong \dfrac{Fh}{L} \end{array} \right\} \tag{2.52}$$

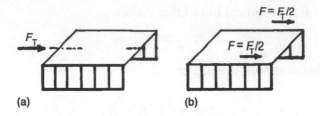

(a)   (b)

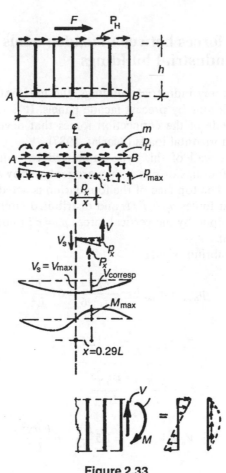

Figure 2.33

☐ **Numerical example 2.8**

$F = 1120$ kN. Wall length 26 m; height 12 m; thickness 0.16 m.

$$M_{max} = 0.1 \times 1120 \times 12 = 1344 \text{ kN m}$$

In section:

$$\bar{x} = 0.29 \times 26 = 7.54 \text{ m}$$

In the same section:

$$V_{\text{corresp}} = \frac{Fh}{L} = \frac{1120 \times 12}{26} = 517 \text{ kN}$$

Section modulus:

$$\frac{0.16 \times 12^2}{6} = 3.8 \text{ m}^3$$

Maximum stress:

$$\sigma_{\text{max}} = \frac{1344}{3.8} = 354 \text{ kN m}^{-2} \cong 0.35 \text{ MPa}$$

Let us assume that the panels are connected at distances of $a = 2$ m. Each connection is subjected to:

Horizontal forces:

$$P_H \lessgtr \sigma_{\text{max}} ba = 354 \times 0.16 \times 2.00 = 112 \text{ kN}$$

Vertical shear stresses:

$$\tau_{\text{max}} = \frac{V}{bz} \cong \frac{517}{0.16 \times (0.85 \times 12)} = 317 \text{ kN m}^{-2} \cong 0.32 \text{ MPa}$$

Vertical shear forces:

$$P \leqslant 317 \times 0.16 \times 2.00 = 101 \text{ kN}$$

On the axis of symmetry:

$$M_s = 0; \ V_s = 1.5F\frac{h}{L} = 1.5 \times 1120 \times \frac{12}{26} = 775 \text{ kN}$$

$$\tau_{\text{max}} = \frac{775}{0.16(0.85 \times 12)} = 475 \text{ kN m}^{-2} \cong 0.48 \text{ MPa}$$

$$P_H = 0; \ P_v = 475 \times 0.16 \times 2.00 = 152 \text{ kN}$$

Assuming foundations at distances of 5.2 m, these are acted upon by the forces:

$$p_{\text{max}} = \frac{6Fh}{L^2} = \frac{6 \times 1120 \times 12}{26^2} = 119.3 \text{ kN m}^{-1}$$

Reactions:

$$R_v \lessgtr 119.3 \times 5.2 = 620 \text{ kN}; \qquad R_H \lessgtr \frac{1120}{6} = 190 \text{ kN} \qquad \square$$

Bibliography

Anon (1982) *The 1977 March 4 Earthquake in Romania*, Ed. Acad. RSR (in Romanian).

Aoyama, H. (1981) A method for the evaluation of the seismic capacity of existing RC buildings in Japan. Bulletin of the NZ National Society for Earthquake Engineering **14**(3), 105–130.

Arcan, M. (1964) Berechnungsverfahren fur Wandscheiben mit einer Reihe Offnungen. *Die Bautechnik*, **41**(3), 95–100.

Beck, H. (1962) Contribution to the analysis of coupled shear walls. *Journal of the American Concrete Institute*, **59**, 1055–1070.

Clough, R., Gulkan, P., Mayers, R. and Carter, E. (1990) Seismic testing of single-story masonry houses, Pt 2. *Journal of Structural Engineering ASCE*, **116**, 257–274.

Dunkerley, S. (1895) On the whirling and vibration of shafts. *Philosophical Transactions of the Royal Society*, **185**, 269–360.

Fintel, M. (1991) Shear walls–an answer for seismic resistance? Construction International, **13** (July), 48–53.

Glogau, O. (1980) Low rise RC buildings of limited ductility, *Bulletin of the NZ National Society for Earthquake Engineering*, **13**(2), 182–190.

Govindan, P., Lakshmipaty, M. and Santhakumar, A. (1986) Ductility of infilled frames. *Journal of the American Concrete Institute*, **83**, 567–576.

Gulkan, P., Clough, R., Mages, R. and Carter, E. (1990) Seismic testing of single-storey masonry houses, Pt 1. *Journal of Structural Engineering ASCE*, **1**(116), 235–256.

Hart, G. (1989) Seismic design of masonry structures, in *The Seismic Design Handbook* (ed F. Naeim), Van Nostrand, New York, Ch. 10.

Hendry, A. (1990) *Structural masonry*, Macmillan.

McLeod, I. (1971) *Shear Wall-Frame Interaction*, Portland Cement Association, Skokie, IL.

Paulay, T. and Priestley, M. (1992) *Seismic Design of Reinforced Concrete and Masonry Buildings*, J. Wiley & Sons, New York.

Paulay, T. and Taylor, R. (1981) Slab coupling of earthquake-resisting shear walls, *Journal of the American Concrete Institute*, **83**, 130–140.

Penelis, G. (1969) Eine Verbesserung der R. Rosman–H. Beck Methode. *Der Bauingenieur*, **44**(12), 454–458.

Rosman, R. (1966) *Tables for Internal Forces of Pierced Shear Walls Subject to Lateral Loads*, W. Ernst & Sohn, Berlin.

Scarlat, A. (1987) *Precast Facade Panels Subject to Seismic forces, Handasah*, December, 11–13 (in Hebrew).

Shiga, T. (1977) Earthquake damage and the amount of walls in RC buildings, in *Proceedings of 6th WCEE*, New Delhi, pp. 2467–2472.

Stafford Smith, B. and Carter, C. (1969) A method of analysis for infilled frames, *Proceedings of the Institution of Civil Engineers, London*, **44**, 31–48.

Stafford Smith, B. and Coull, A. (1991) *Tall Building Structures: Analysis and Design*, J. Wiley & Sons, New York.

Wakabayashi, M. (1986) *Design of Earthquake Resistant Buildings*, McGraw-Hill, New York.

Weaver, W. Jr and Gere, J. (1980) *Matrix Analysis of Framed Structures*, Van Nostrand, New York.

Wood, S. (1991) Performance of RC buildings during the 1985 Chile earthquake: implications for the design of structural walls. *Earthquake Spectra*, 7 (4), 607–638.

Uniform Building Code (UBC) (1991) International Conference of Building Officials, Whittier.

Uniform Code for Building Construction (UCBC) (1992) Appendix, Ch. 1: Seismic provisions for unreinforced masonry bearing walls and buildings (Advance version).

3 Dual systems (structural walls and frames)

3.1 Introduction

Most structures are dual, and we only conventionally refer either to building frame systems (by neglecting the structural walls), or to structural walls systems (by neglecting the columns). Dual systems are mostly taken into consideration up to 30–40 storeys. Above this limit either RC tube structures (see section 4.2) or steel structures are commonly used.

Dual systems are generally space structures but, as can be seen in Figure. 3.1, we may perform the analysis in two stages: a planar and a three-dimensional (torsion) problem. Instead of the resultant lateral force F passing through the **centre of forces** (centre of masses when we deal with seismic forces), denoted as CM in Figure 3.1, we consider an equivalent system of forces: the resultant force F passing through the **centre of rigidity** CR, chosen so that it involves mostly translations and corresponding **translational forces** (i.e. a planar problem), plus the **storey moment of torsion** $M^T = F \cdot e$ (i.e. a spatial problem), yielding additional **torsional forces** (see section 4.1).

Generally, we have to determine the centres of masses as well as the centres of rigidity for each storey separately. In the case of uniform or nearly uniform structures we can assume identical centres of masses and identical centres of rigidity.

In the following we shall assume the slabs are rigid in their planes as compared with the vertical structural walls (the assumption is valid in most cases, although not always).

3.2 Analysis of dual systems: classical approach

3.2.1 ANALYSIS OF DUAL SYSTEMS ON PLANAR SCHEMES

The main problem to be dealt with is the correct distribution of the lateral forces between frames and structural walls, as a result of their interaction. After solving this problem we may return to the approximate methods presented in Chapter 1 (for frames) and Chapter 2 (for structural walls).

Consider the structure shown in Figure 3.2a. As the effect of torsion is neglected at this stage the deflections of each substructure 1, 2,... parallel to the resultant force F are equal. Consequently, the analysis may be performed

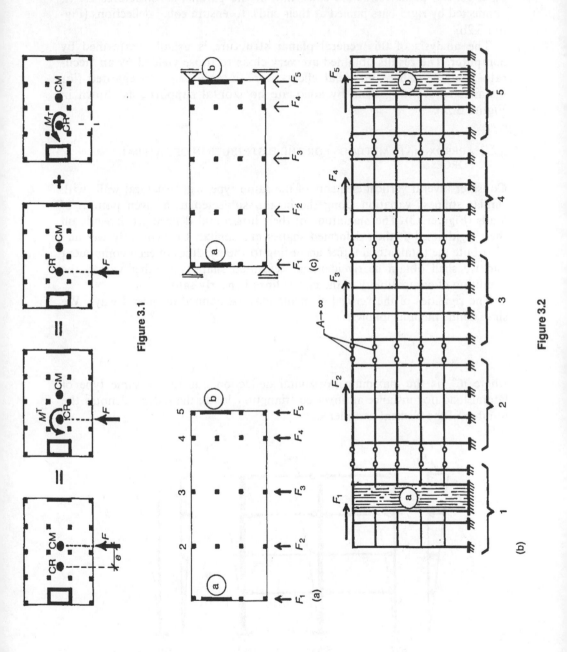

Figure 3.1

Figure 3.2

on a general planar structure consisting of the parallel substructures 1, 2,...
connected by rigid bars pinned at their ends, to ensure equal deflections (Figure 3.2b).

The analysis of this general planar structure is usually performed by
computer. The results obtained are very close to those yielded by an 'accurate analysis', where torsional effects were intentionally disregarded (for
instance: preventing them by adequate horizontal supports, as shown in
Figure 3.2c).

3.2.2 APPROXIMATE METHODS FOR THE DISTRIBUTION OF LATERAL FORCES

Consider several vertical elements of the same type, e.g. structural walls with
similar stiffness variation along the height, subjected to a given pattern of
loads (Figure 3.3). The magnitude of their horizontal deflections depends on
their rigidities, but their deformed shapes are similar. Consequently we may
distribute the total lateral forces according to the rigidities of each component
and we shall obtain nearly identical deformed shapes: in other words, the
condition of compatibility of the elastic lines is nearly satisfied.

The rigidities of the vertical elements may be defined in several ways. We
shall define them as follows:

$$k_i = \frac{1}{u_{i_{max}}}$$

where $u_{i_{max}}$ is the maximum horizontal deflection due to the same type of
lateral load (for instance an inverted triangular load); the index i denotes the
number of the element considered.

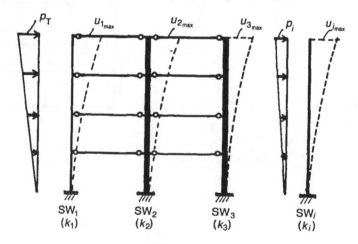

Figure 3.3

The coefficient of distribution is defined as

$$d_i = \frac{k_i}{\sum k} \tag{3.1}$$

We point out that the coefficient of distribution d_i does not significantly differ if we take into consideration another type of lateral loading in defining the rigidity k_i (obviously, on condition that we use the same loading for all the vertical elements).

Each substructure is acted upon by the lateral forces:

$$p_i = p_T d_i = \frac{p_T k_i}{\sum k} \tag{3.2}$$

In the case of dual systems (Figure 3.4) the problem of distribution of lateral loads is more complex. We shall distinguish between two types of dual system: structural walls with openings and frames and structural walls without openings and frames.

(a) Structural walls with openings and frames

It will be recalled that coupled structural walls are an intermediate structure between structural walls without openings and frames. When the openings are large (doors or large windows) their deformed shape is close to that of frames. In such cases we can distribute the lateral loads in proportion to the rigidities of each vertical element (equation (3.2)).

(b) Structural walls without openings and frames

When the structural walls do not have openings or when the openings are small, then their deformed shape significantly differs from that of frames, and we have to resort to more sophisticated techniques (Figure 3.5). We shall use the concept of **equivalent dual structure** proposed by Khan and Sbarounis (1964).

We replace the frames by an equivalent one-bay frame, as defined in section 1.2:

$$I_c^* = \frac{\sum I_c}{2} \quad \text{(columns)}; \quad I_b^* = \sum I_b \quad \text{(beams)}$$

In the same way, we replace the structural walls by an equivalent structural wall, defined as

$$I_{sw}^* = \sum I_{sw}; \quad A_{sw}^* = \sum A_{sw}$$

The equivalent mixed structure is made up by the equivalent frame connected to the equivalent structural wall by the usual connection bars meant to ensure equal horizontal deflections (Figure 3.6).

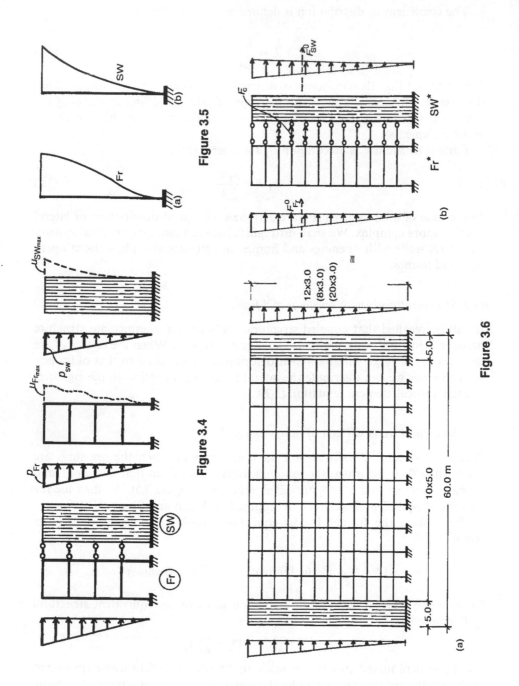

Figure 3.5

Figure 3.4

Figure 3.6

We let, based on numerical examples:

$$u^* \cong u; \qquad \sum M_c^* = 2M_c^* \cong \sum M_c; \qquad M_{sw}^* \cong \sum M_{sw} \qquad (3.3)$$

Finally, we distribute $M_c^* = 2M_c^*$ (M_{sw}^*) to each column (structural wall) according to their rigidities.

It is usual to distribute the resultant stresses between columns and structural walls according to their rigidities, computed for each storey. This procedure may lead to significant errors due to the different effects of shear forces on the considered rigidities. For squat structural walls $(H < l/5)$ we have to consider the effect of the shear forces on the rigidity by using the data given in section 2.2.1. Numerical computations show that an acceptable decrease in the rigidities of the structural walls can be obtained by multiplying them by the coefficient:

$$\frac{1}{1 + 2s} \qquad (3.4)$$

where s is defined by equation (2.2a):

$$s = \frac{6f\,EI}{GAH^2}$$

f is the **shape factor** of the transversal section.
In the case of rectangular sections:

$$s = \frac{1 \cdot 41\,l^2}{H^2}$$

We point out that H denotes the total height of the structure. We have computed uniform equivalent structures with 8, 12 and 20 storeys with $3 \cdot 0$ m storey height by taking into account three rigidity ratios $(k_{sw}/k_{Fr} = \frac{1}{4}; 1; 4)$; for each case, lintels with sections $0 \cdot 20 \times 0 \cdot 50$ m and $1 \cdot 00 \times 0 \cdot 20$ m (slab beams) have been considered.

The structures have been loaded with inverted triangular and uniformly distributed loads. Full fixity at base has been assumed (Figure 3.6).

A typical variation of the interaction forces F_c between the equivalent frame and the equivalent structural wall is displayed in Figure 3.7; it has been drawn assuming that the given loads have been distributed to the component structures in proportion to their rigidities.

The curves F_{sw}/F_T and M_{sw}/M_T are displayed in Figure 3.8, where $F_{sw} = F_{sw}^o + F_c$ and M_{sw} are the resultant forces and the overturning moment acting on the equivalent structural wall; F_T and M_T are the resultant force and the overturning moment of the given loads acting on the entire structure.

The curves represent average values obtained for the two types of loading mentioned above.

Notes: (a) Most of the lateral loads F_T are taken by the structural wall, even when its rigidity is relatively low. Consequently the bending moments occurring in the frames are rather small.

When the structural walls are much more rigid than the frames ($k_{sw}/k_{Fr} \gtrless 6$) we may consider that the horizontal forces are taken by structural walls only, and the frames are 'braced'.

(b) The overturning moment acting on the structural walls (M_{sw}) varies between 30% of the total overturning moment M_T (for $k_{sw}/k_{Fr} = \frac{1}{4}$) and 80% (for $k_{sw}/k_{Fr} = 4$).

Note that these results are based on the fixed supports assumption. The results obtained by considering elastic supports, as shown in section 3.3, are quite different.

(c) Consider a structure made up of several complex substructures $S_1, S_2, ... S_n$; we assume that the total number of unknowns is excessive.

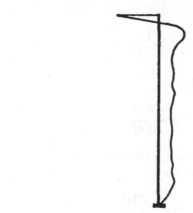

Figure 3.7

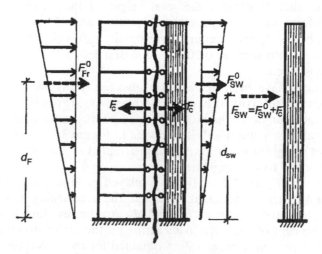

Figure 3.8

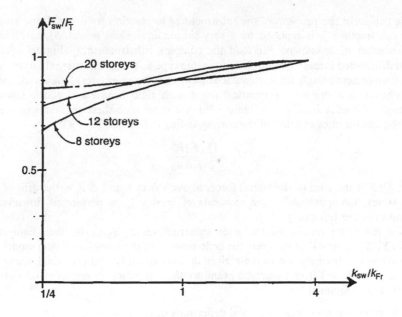

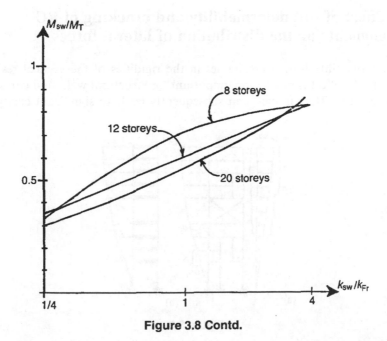

Figure 3.8 Contd.

We can solve the problem of the substructures interaction by using **elastic models**. Each substructure S_i is replaced by a very simple equivalent model, S_i', involving a small number of unknowns. We load the complex substructure S_i (Figure 3.9a) by lateral distributed forces – preferably of the same type as the given external forces – and we determine accordingly the horizontal deflections u_i (by computer). The elastic model S_i' is chosen as a one-bay symmetrical frame with rigid beams and elastic columns (moments of inertia $I_{c1}', I_{c2}', \ldots$) (Figure 3.9b). We determine these moments of inertia from the condition of equality of the corresponding deflections ($u_i' = u_i$):

$$I_{c_i}' = \frac{(\sum_i F)h_i^3}{24E\,\Delta u_i} \tag{3.5}$$

where $\sum_i F$ is the sum of the lateral forces above storey i and Δu_i is the drift of the same storey. Computation of the moments of inertia I_{c_i}' is performed downwards, starting from the top storey.

As a result, we replace all the given substructures $S_1, S_2, \ldots$ by their equivalent models $S_1', S_2'. \ldots$ and then compute the deflections u_i' of the complete elastic model; we load each real substructure with these given deflections and obtain the final stresses.

It has been found from numerical examples that statisfactory results (errors of less than 10%) are obtained for:

- the computation of the horizontal deflections of the structure;
- the first mode of vibration and the corresponding fundamental period.

The resultant stresses are satisfactory only when the structural walls are more rigid than the frames ($k_{sw} \gtrsim k_{Fr}$).

3.3 Effect of soil deformability and cracking of RC elements on the distribution of lateral forces

Soil deformability leads to decreases in the rigidities of the vertical resistant elements; the effect is much more important for structural walls and cores than for moment-resisting frames, and subsequently leads to significant changes in

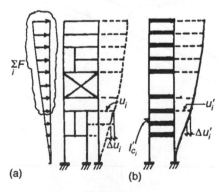

Figure 3.9

the distribution of lateral forces with respect to the results yielded by a classical analysis, where this effect is disregarded (Scarlat, 1993).

In order to assess the order of magnitude of this effect we have analysed the dual systems shown in Figure 3.6, on the assumption that the supports are elastic, and we have compared the results with those obtained by considering fixed supports (Figure 3.10: bending moments acting on the shear wall).

The bending moment diagram acting on the structural wall yielded by a fixed supports analysis is displayed in Figure 3.10a; it is close to a cantilever diagram (the positive moments usually remain less than 10% of the base moments M_f°).

The diagram displayed in Figure 3.10b is based on the assumption of a deformable soil with a high subgrade modulus ($k_s = 100\,000\,\text{kNm}^{-3}$); the base moment M_f' decreases to nearly half of the fixed support moment M_f° and the positive moment increases correspondingly.

The diagram displayed in Figure 3.10c corresponds to a low subgrade modulus ($k_s = 20\,000\,\text{kNm}^{-3}$). The base moment ($M_f''$) decreases to $\frac{1}{5}-\frac{1}{3}$ of the fixed end moment M_f°, leading to significant positive moments; the moment diagram is close to that of a simply supported column.

In order to assess the order of magnitude of the changes involved by considering the soil deformability we shall refer to the eight-storey dual structure shown in Figure 3.11. Four analyses have been performed, as follows.

(a) Fixed supports

The lateral forces are distributed between the structural wall and the frame in proportion with their rigidities. By loading them separately with the same uniformly distributed loads we obtain

$$u_{Fr_{max}} = 2\cdot667; \qquad k_{Fr} = \frac{1}{2\cdot667} = 0\cdot375$$

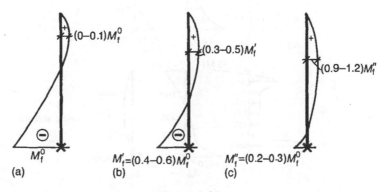

Figure 3.10

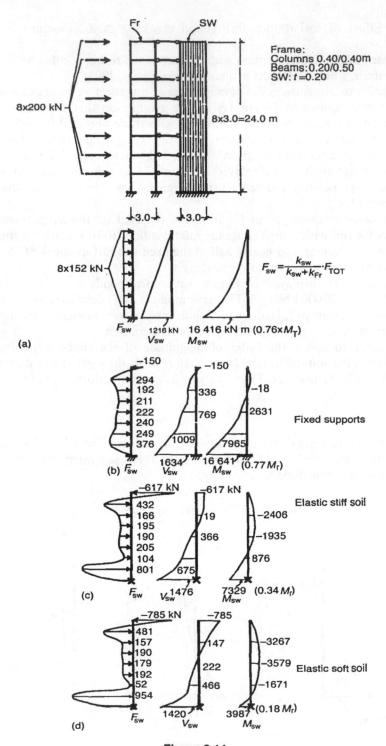

Frame:
Columns 0.40/0.40m
Beams: 0.20/0.50
SW: t = 0.20

8x200 kN

8x3.0=24.0 m

$\downarrow$3.0$\downarrow$ $\downarrow$3.0$\downarrow$

8x152 kN

$F_{SW} = \dfrac{k_{SW}}{k_{SW}+k_{Fr}} \cdot F_{TOT}$

F_{SW} 1216 kN   16 416 kN m (0.76xM_T)
          V_{SW} M_{SW}

(a)

−150   −150
294
192   336   −18
211
222   769   2631
240
249   1009   Fixed supports
376
         1634   7965
(b) F_{SW} 16 641 (0.77 M_r)
         V_{SW} M_{SW}

−617 kN   −617 kN
432
166   19   −2406   Elastic stiff soil
195
190   366   −1935
205
104   876
801   675
(c) F_{SW} V_{SW} 1476   7329 (0.34 M_r)
                          M_{SW}

−785 kN   −785
481
157   147   −3267
190
179   222   −3579   Elastic soft soil
192
52   466   −1671
954
(d) F_{SW} 1420   3987 (0.18 M_r)
              V_{SW} M_{SW}

Figure 3.11

$$u_{sw_{max}} = 0.845; \qquad k_{sw} = \frac{1}{0.845} = 1.18 = 3.15 k_{Fr}$$

$$\frac{k_{sw}}{k_{sw} + k_{Fr}} = 0.76$$

(no interaction forces were taken into consideration).

The corresponding diagrams of the shear forces V_{sw} and bending moments M_{sw} acting on the structural wall are displayed in Figure 3.11a.

(b) Fixed supports

The lateral loads act on the dual system and an accurate interaction analysis is performed. The total forces F_{sw} acting on the structural wall (given loads + interaction forces) and the corresponding diagrams V_{sw} and M_{sw} are displayed in Figure 3.11b. The maximum deflection $u_0 = 1.04$.

(c) Elastic supports

The analysis of point (b) has been repeated by assuming a stiff deformable soil (subgrade modulus $k_s = 100\,000\,\text{kNm}^{-3}$). The results are displayed in Figure 3.11c. The maximum deflection $u' = 3.24$.

(d) Elastic supports

By assuming a soft deformable soil ($k_s = 20\,00\,\text{kNm}^{-3}$); we obtain the results displayed in Figure 3.11d. The maximum deflection $u'' = 4.03$.

The moments acting on the foundations–points (c) and (d)–are compared with the total overturning moment due to the lateral loads ($M_T = 19\,200\,\text{kNm}$). We see that the results (diagrams V_{sw}, M_{sw}) and the maximum deflections obtained by taking into account the soil deformability differ sharply from the corresponding results yielded by the usual analysis, where fixed supports are assumed. As the data dealing with soil deformability are often not reliable, particularly when cyclic dynamic loads occur (as during earthquake attacks), the reliability of the results based on analyses ignoring this effect is questionable, too.

Additional doubts arise when we refer to the effect of cracking and inelastic behaviour of RC elements. As shown in Appendix A3, the order of magnitude of the decrease in rigidity of structural elements due to this effect varies between 30% and 80%, depending on the type of element, the existing reinforcement, and the intensity of compressive stresses.

It is therefore legitimate to question the validity of the 'classical analysis' of dual systems we perform. The author's opinion is that the picture provided by such analyses is strongly distorted, and we have to look to other ways of analysing dual systems.

In the following we shall refer to two approaches when dealing with the structural analysis of dual systems: a classical approach and a limit design approach.

The **classical approach** is based on two main simplifying assumptions:

- the structural elements are considered in the elastic range and the reinforced concrete elements as uncracked;
- the foundation soil is rigid (we neglect the effect of the soil's deformability).

The distribution of the horizontal forces between structural walls and columns is performed in agreement with an elastic analysis, either spatial or planar or a combination of both. A 'complete spatial analysis' gives sometimes the agreeable illusion of an 'accurate analysis'.

This approach is recommended in most modern codes, although many engineers are well acquainted with its shortcomings. Both assumptions disregard the significant decrease in rigidity as a result of each of the aforementioned assumptions. What is more, the decrease in rigidity sharply differs as a function of type of the structural members (see Appendix C). However, the temptation of the simplicity is yet too great to be resisted. Several amendments are recommended in order to take into account the temporary excursions of at least several structural members into the inelastic range (see Chapter 6).

In the **limit design approach**, the total resisting force results as a sum of the resisting forces of the component elements in each direction; in order to take into account the different ductilities of each type of structural element the forces are multiplied by various **participation factors**. Such an approach is presently used in the frame of 'first screening' procedures. The author is aware of only one seismic code permitting such a technique in structural analysis: the Japanese code (1987).

Bibliography

Anastasiadis, K. and Avramidis, I. (1988) Einheitliche Methode fur die Berechnung gekoppelter Rahmen-Scheiben System auf elastischer Grundung. *Bautechnik*, **65**, (4).

Khan, F. and Sbarounis, G. (1964) Interaction of shear walls and frames. *Journal of Structural Division ASCE*, **ST3**, 285–336.

McLeod, I. (1971) *Shear Wall-Frame Interaction*, Portland Cement Association, Skokie, IL.

Paulay, T. and Priestley, M. (1992) *Seismic Design of Reinforced Concrete and Masonry Buildings*, J. Wiley & Sons, New York.

Scarlat A. (1989) Critères sismiques dans le project du bâtiment Shikmona, Haifa, in *2ème Colloque National AFPS*, St Rémy, pp. CB/2 7–15.

Scarlat, A. (1993) Soil deformability effect on rigidity-related aspects of multistory buildings analysis, *ACI Str. Journal*, **90** (2) 156–162.

Taranath, B. (1988) *Structural Analysis and Design of Tall Buildings*, McGraw-Hill, New York.

4 Space structures

4.1 Torsional forces: Lin's theory

Consider an asymmetric multi-storey structure subjected to lateral loads due to earthquake or wind; the forces are parallel to a given direction, for instance to the axis Y (Figure 4.1). We assume that the slabs are rigid in both the horizontal and vertical planes; we neglect the axial deformations of the vertical resisting elements. We have to point out that the assumption of the rigidity of the slabs in the vertical plane is obviously untrue in most cases; it entails 'anomalies', which are further detailed.

The aforementioned assumptions enable us to check the displacements and the stresses of columns and structural walls separately for each storey (by 'decoupling' the unknowns of the problem).

At first we assume that the storeys are identical (Figure 4.2). Consider the storey j of the structure. $V_j = \sum_j^n F_j$ denotes the sum of the forces above the considered storey (the **storey shear force**).

The relative displacement of the slabs $j + 1$ and j can be resolved into two components:

- a translation parallel to the given forces, accompanied by 'translational forces and stresses';
- a rotation about a vertical axis passing through the **centre of rigidity** CR, accompanied by additional 'torsional forces and stresses'.

The position of the resultant force V_j depends on the type of forces: in the case of wind pressure, at mid-length ($L/2$); in the case of seismic forces, V_j passes through the **centre of mass** CM. In determining the position of the centre of mass we have to take into account all existing elements of the storey j, including non-structural ones. In most cases the centre of mass is close to the centre of gravity of the slab, considered as a geometrical area, and we can replace the centre of mass by the centre of gravity of the slab's area.

As the storeys are identical, the positions of the centres of rigidity CR are the same; also, the positions of the centres of mass are the same for all storeys.

The distance e between the centre of rigidity and the direction of the resultant force V_j is referred to as the **eccentricity of the resultant force** V_j. The **storey torsion moment** acting on the storey j results in

$$M^\mathrm{T} = V_j e \tag{4.1}$$

In the case of non-uniform storeys (Figure 4.3) the position of the centres of rigidity CR, as well as the positions of the centres of mass CM, differ from storey to storey.

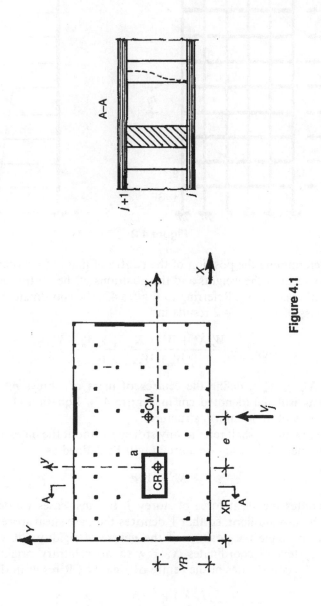

Figure 4.1

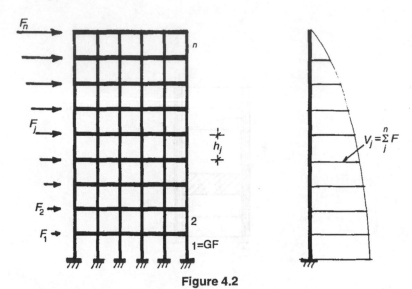

Figure 4.2

When determining the position of the centre of mass of a storey j, we have to take into account the weights and the positions of the centres of mass of all the storeys above storey j. Referring to Figure 4.3, the coordinate $X_{\text{CM}_{(n-2)}}$ of the centre of mass of storey $n-2$ results in

$$X_{\text{CM}_{(n-2)}} = \frac{W_n X'_n + W_{n-1} X'_{n-1} + W_{n-2} X'_{n-2}}{W_n + W_{n-1} + W_{n-2}}. \tag{4.2}$$

where X'_n, X'_{n-1}, X'_{n-2} define the centres of mass of storeys n, $n-1$, $n-2$, considered as isolated (denoted cm in Figure 4.3). Equation (4.1) holds true also in the case of non-uniform structures.

In the following we shall refer to any storey j without the index j. The storey torsional moment M^T acting on storey j will be defined as

$$M^\text{T} = V/e \tag{4.1a}$$

where V denotes the shear force of storey j. In most cases we deal with the torsion at the ground floor, so that V denotes the base shear force.

In order to define the position of the centre of rigidity CR we choose a temporary system of coordinates X, Y with an arbitrary origin. It can be shown that the coordinates of the centre of rigidity CR result in (Figure 4.1):

$$\left. \begin{aligned} XR &= \frac{\Sigma(I_{x_i} X_i) + I_{x_a} \cdot X_a + I_{x_b} X_b}{\Sigma I_{x_i} + I_{x_a} + I_{x_b}} \\[2mm] YR &= \frac{\Sigma(I_{y_i} Y_i) + I_{y_a} \cdot Y_a + I_{y_b} Y_b}{\Sigma I_{y_i} + I_{y_a} + I_{y_b}} \end{aligned} \right\} \tag{4.2}$$

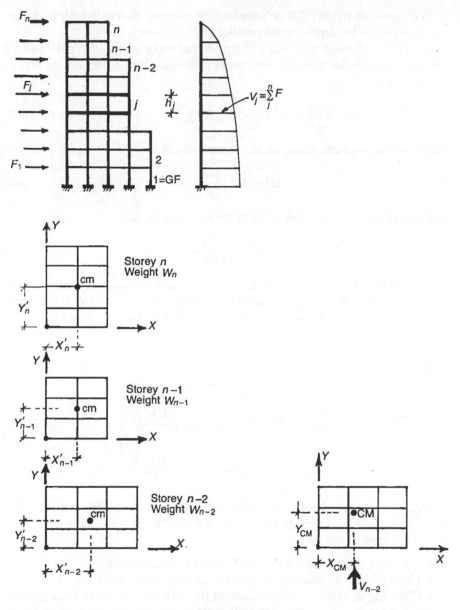

Figure 4.3

where I_{x_i}, I_{y_i} are the moments of inertia of column i. For the definition of the moments of inertia of the structural walls (I_{x_a}, I_{y_a}, ...), see Note (a) below.

After determining the position of CR, we use the final system of coordinates x, y with the origin in CR.

The centre of rigidity CR is 'attracted' by vertical elements with high moments of inertia (structural walls parallel to F or cores).

Lin (1951) showed that when the principal axes of the vertical resisting elements are parallel, then the additional torsional shear forces acting on the element i are

$$\Delta V_{x_i}^{\mathrm{T}} = M^{\mathrm{T}} y_i \frac{I_{y_i}}{C_0}; \qquad \Delta V_{y_i}^{\mathrm{T}} = M^{\mathrm{T}} x_i \frac{I_{x_i}}{C_0} \tag{4.3}$$

The torsional moment acting on the same element is

$$M_i^{\mathrm{T}} = M^{\mathrm{T}} \frac{G h^2}{12 E C_0} I_{t_i} \tag{4.4}$$

The slab describes a rotation in its plane with the angle

$$\varphi^{\mathrm{T}} = \frac{M^{\mathrm{T}} h^3}{12 E C_0} \tag{4.5}$$

where h is the storey height (m); I_{t_i} is the torsional moment of inertia of the element i (m^4); E is the Young modulus of elasticity (kN m^{-2}); G is the shear modulus of elasticity (kN m^{-2}); C_0 is the torsion constant (m^6); and

$$C_0 = C_{\mathrm{col}} + C_{\mathrm{a}} + C_{\mathrm{b}}$$

where

$$\left. \begin{array}{l} C_{\mathrm{col}} = \sum (I_{x_i} x_i^2) + \sum (I_{y_i} y_i^2) \\[2mm] C_{\mathrm{a}} = I_{x_a} x_{\mathrm{a}} + I_{y_a} y_{\mathrm{a}} + \dfrac{G h^2 I_{t_a}}{12 E} \\[3mm] C_{\mathrm{b}} = I_{x_b} x_{\mathrm{b}} + I_{y_b} y_{\mathrm{b}} + \dfrac{G h^2 I_{t_b}}{12 E} \end{array} \right\} \tag{4.6}$$

According to equations (4.3), the torsional shear forces increase with the distance to the centre of rigidity CR. Figure 4.4 shows the torsional moments of inertia for several sections:

- Cases (b), (c), (d), (e): Thin-walled sections are assumed.
- Cases (c), (e): A_0 denotes the area inside the axes of the wall.
- Cases (d): In this case the presence of the interior walls may be neglected.
- Cases (e): In the case of cores with openings an equivalent thin wall with a thickness t_{eq} may be considered; the torsional moment of inertia I_t will be computed as in case (c); f denotes the **shape factor** (see section 2.2.1). The expression of the equivalent thickness t_{eq} has been evaluated by Khan and Stafford Smith (1975). Usually, the thickness t_{eq} is in the range 5–30 mm. It leads to a very significant decrease in the torsional moment of inertia,

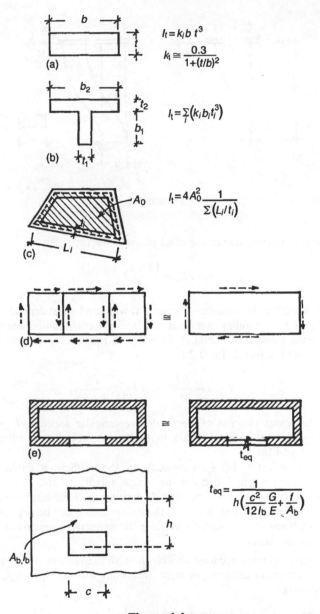

$$I_t = k_i b \, t^3$$

$$k_i \cong \frac{0.3}{1+(t/b)^2}$$

(a)

$$I_t = \sum_i \left(k_i b_i t_i^3\right)$$

(b)

$$I_t = 4 A_0^2 \frac{1}{\sum \left(L_i / t_i\right)}$$

(c)

(d)

(e)

$$t_{eq} = \frac{1}{h \left(\dfrac{c^2}{12 I_b} \dfrac{G}{E} + \dfrac{f}{A_b} \right)}$$

Figure 4.4

which may reach 50–80% for small cores and 30–70% for large cores.

Notes: (a) The deformation of the columns is mainly moment dependent (Figure 4.5); consequently their lateral rigidity may be defined as a function of their moments of

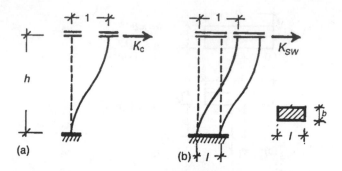

Figure 4.5

inertia without correction due to the effect of shear forces:

$$K_c = \frac{12\,E\,I_c}{h^3} \tag{4.7}$$

On the other hand, the deformation of structural walls and cores depends, in most cases on the effect of both bending moments and shear forces; consequently we have to correct the usual moment of inertia I_{sw}° in order to account for the effect of the shear deformations (see Chapter 2, Table 2.1):

$$K_{sw} = \frac{12\,E\,I_{sw}}{h^3}; \qquad I_{sw} \cong \frac{I_{sw}^{\circ}}{1+2s} \tag{4.8}$$

I_{sw}° denotes the usual moment of inertia (for rectangular sections, $I_{sw}^{\circ} = bl^3/12$). I_{sw} denotes the corrected moment of inertia, by taking into account the effect of the shear forces, too (Figure 4.5b).

The 'correction factor' $1/(1+2s)$ depends on the coefficients s (see section 2.1.1): $s = 6fE\,I_{sw}^{\circ}/(G\,A\,h^2)$, where f denotes the 'shape factor'. In the case of rectangular sections, $s = 1\cdot41\,l^2/H^2$. We point out that H denotes the total height of the structure. When $H/l > 5$ the effect of shear forces on the deformations can be neglected ($I_{sw} \cong I_{sw}^{\circ}$). In equations (4.2) and (4.6) we have to consider the corrected moments of inertia I_{sw} for structural walls and cores.

(b) Lin's theory gives more accurate results when all the vertical resisting elements are of the same type: either columns, or structural walls without openings, or structural walls with openings.

(c) The torsional rigidity of usual columns is negligible.

(d) The torsional rigidity of thin-walled closed sections is much higher than that of thin-walled open sections (we define a thin-walled section when the thickness t of any component wall is less than 1/10 of its length). In the case of dual structures when open section resisting elements are predominant, very high shear and additional (warping) normal stresses may develop. The coordinates X, Y and x, y of thin-walled open sections refer to the **shear centres** SC of these sections rather than to their centres of

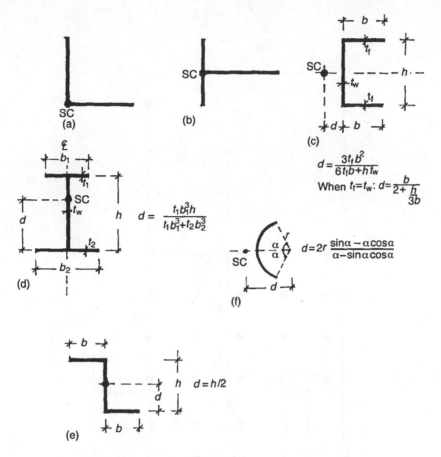

Figure 4.6

gravity. Figure 4.6 gives the positions of the shear centres for several thin-walled open sections.

(e) Owing to the uncertainty as to the mass distribution, a minimum eccentricity must be considered even for symmetric structures. Most codes (e.g. the UBC-92 code) recommend a value of $0 \cdot 05 L$ (L denotes the length of the slab, normal to the force V).

(f) Owing to its basic assumptions, Lin's theory is essentially an approximate one. When we relinquish these assumptions, the analysis becomes an 'accurate' one; in this case several definitions are possible for the centre of rigidity of a given storey (Jiang et al., 1993), but their practical interest is limited.

□ **Numerical example 4.1**

Consider the eight-storey building with identical slabs, shown in Figure 4.7 (total height $H = 8 \times 3 \text{ m} = 24 \text{ m}$). Zone I: columns $0 \cdot 80/0 \cdot 30$ m. Zone II: columns $0 \cdot 50/0 \cdot 50$ m. RC structural walls a and b: $0 \cdot 20/5 \cdot 00$ m.

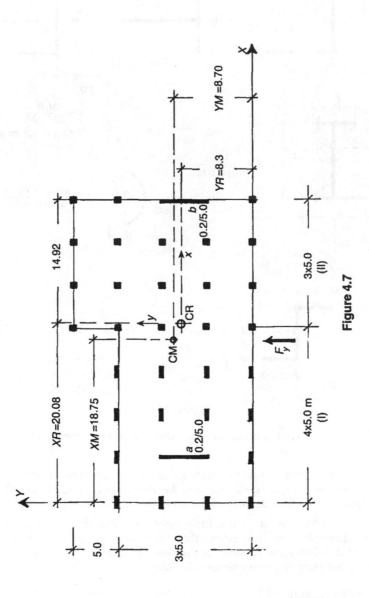

Figure 4.7

We assume that the centre of mass CM coincides with the centre of gravity of the slab's area:

$$XM = 18{\cdot}75\,\text{m}; \qquad YM = 8{\cdot}70\,\text{m}$$

We shall determine the torsional forces in the vertical elements of the ground floor according to Lin's theory, by assuming non-deformable soil (fixed ends).

We consider a concentrated force $F = F_y = 1000\,\text{kN}$ acting in the centre of mass of the top slab. The shear force is constant along the height. The base shear (at ground floor) is $V = 1000\,\text{kN}$. Columns in zone I: $I_{x_i} = 0{\cdot}0018\,\text{m}^4$; $I_{y_i} = 0{\cdot}0128\,\text{m}^4$. Columns in zone II: $I_{x_i} = I_{y_i} = 0{\cdot}00521\,\text{m}^4$.

$$\sum I_{x_i} = 14 \times 0{\cdot}0018 + 18 \times 0{\cdot}00521 = 0{\cdot}119\,\text{m}^4$$

$$\sum I_{y_i} = 14 \times 0{\cdot}0128 + 18 \times 0{\cdot}00521 = 0{\cdot}273\,\text{m}^4$$

$$\sum (I_{x_i} X_i) = 0{\cdot}0018\,(4 \times 0 + 2 \times 5 + 4 \times 10 + 4 + 15)$$
$$+ 0{\cdot}00521\,[5(20 + 25 + 30) + 3 \times 35] = 2{\cdot}698$$

$$\sum (I_{y_i} Y_i) = \cdots = 2{\cdot}307$$

Structural walls a and b:

Moments of inertia about x:$I_x^\circ = \dfrac{0{\cdot}2 \times 5^3}{12} = 2{\cdot}083$

As $H/l = 24/5 = 4{\cdot}80 < 5$ we shall consider the effect of shear forces on the deformations. The correction factor:

$$s = \frac{1{\cdot}41\,l^2}{H^2} = \frac{1{\cdot}41 \times 5^2}{24^2} = 0{\cdot}0612; \qquad \frac{1}{1 + 2s} = 0{\cdot}891$$

$$I_x = 2{\cdot}083 \times 0{\cdot}891 = 1{\cdot}856\,\text{m}^4$$

Moments of inertia about y: As $H/l = 24/0{\cdot}2 = 120 > 5$, we shall neglect the effect of shear deformations:

$$I_y \cong I_y^\circ = \frac{5 \times 0{\cdot}2^3}{12} = 0{\cdot}0033\,\text{m}^4$$

The position of the centre of rigidity CR:

$$XR = \frac{\sum (I_{x_i} X_i) + (I_{x_a} X_a) + (I_{x_b} X_b)}{\sum I_{x_i} + I_{x_a} + I_{x_b}}$$
$$= \frac{2{\cdot}698 + 1{\cdot}856 \times (5 + 35)}{0{\cdot}119 + 2 \times 1{\cdot}856} = 20{\cdot}08\,\text{m}$$

$$YR = \cdots = 8{\cdot}43\,\text{m}$$

The torsional moment of inertia:

$$I_{t_a} = I_{t_b} = \frac{5{\cdot}0 \times 0{\cdot}2^3}{3} = 0{\cdot}0133\,\text{m}^4$$

We let $G/E = 0.425$.

The torsion constant:

$$C_0 = C_{col} + C_a + C_b$$

$$C_{col} = \sum(I_{x_i} x_i^2) + \sum(I_{y_i} y_i^2)$$

$$\begin{aligned}\sum(I_{x_i} x_i^2) &= 0.0018\,[4(20.08 - 0)^2 + 2(20.08 - 5)^2 \\ &\quad + 4(20.08 - 10)^2 + 4(20.08 - 15)^2] + 0.00521\,[5(20.08 - 20)^2 \\ &\quad + 5(20.08 - 25)^2 + 5(20.08 - 30)^2 + 3(20.08 - 35)^2] = 11.31\,m^6\end{aligned}$$

$$\sum(I_{y_i} y_i^2) = \cdots = 11.79\,m^6$$

$$C_{col} = 11.31 + 11.79 = 23.10\,m^6$$

$$C_a = I_{x_a} x_a^2 + I_{y_a} y_a^2 + Gh^2\frac{I_{t_a}}{12\,E} = 1.856\,(20.08 - 5)^2$$

$$+ 0.0033\,(8.43 - 7.5)^2 + \frac{0.425 \times 3^2 \times 0.0133}{12} = 422.1\,m^6$$

$$C_b = \cdots = 413.2\,m^6$$

$$C_0 = 23.10 + 422.1 + 413.2 \cong 858\,m^6$$

The storey torsional moment results in

$$M^T = 1000\,(20.08 - 18.75) = 1330\,kN\,m$$

The torsional shear forces resisted by the structural wall a are

$$\Delta V^T_{y_a} = \frac{M^T I^T_{x_a} x_a}{C_0} = \frac{1330 \times 1.856\,(20.08 - 5)}{858} = 43.4\,kN$$

$$\Delta V^T_{x_a} \cong 0.$$

The torsional moment resisted by the structural wall a is:

$$M_{t_a} = \frac{M^T Gh^2 I_{t_a}}{12\,E\,C_0}$$

$$= 1330 \times 0.425 \times 3^2 \times \frac{0.1333}{12 \times 858} = 0.007 \cong 0$$

The torsional forces taken by the columns are negligible.

An accurate analysis (as a tri-dimensional structure) yields:

$$\text{Eccentricity } e = (XR - XM) = 20.11 - 18.75 = 1.36\,m$$

The storey torsional moment $M^T = 1000 \times 1.36 = 1360\,kN\,m$ ☐

4.2 Approximate torsional analysis of dual structures based on Lin's theory

Dual structures with eccentric structural walls/cores are subjected to high supplementary torsional shear forces. When determining the position of the centre of rigidity CR (the coordinates XR, YR) and the torsional shear forces, the contribution of columns is usually much smaller than the contribution of structural walls/cores. Consequently, we may admit some simplifications regarding the columns without adversely affecting the results (Scarlat, 1986). These simplifications are intended to avoid the cumbersome computations resulting from Lin's theory.

Consider a slab supported on the columns i and the structural walls/cores a, b, ... (Figure 4.8a).

We denote

$$I_x^{col} = \sum I_{x_i}; \qquad I_y^{col} = \sum I_{y_i} \tag{4.9}$$

We replace the columns i (area A_i, moments of inertia I_{x_i}, I_{y_i}) by an infinite number of identical infinitesimal columns (dA, dI_x, dI_y) covering the entire area of the slab (Figure 4.8b). Assuming the area of the slab to be

$$A^* = \int dA$$

we admit:

$$\frac{dI_x}{dA} \cong constant = \frac{I_x^{col}}{A^*}$$

$$\frac{dI_y}{dA} \cong constant = \frac{I_y^{col}}{A^*}$$

It follows that

$$\sum I_{x_i} X_i \cong \int \left(\frac{dI_x}{dA}\right) dA\, X + I_{x_a} X_a + I_{x_b} X_b$$

$$= \left(\frac{I_x^{col}}{A^*}\right) \int X\, dA + I_{x_a} X_a + I_{x_b} X_b$$

$$XR = \frac{\sum(I_{x_i} X_i) + I_{x_a} X_a + I_{x_b} X_b}{\sum I_{x_i} + I_{x_a} + I_{x_b}}$$

$$\cong \frac{(I_x^{col}/A^*)\int X\, dA + I_{x_a} X_a + I_{x_b} X_b}{I_x^{col} + I_{x_a} + I_{x_b}}$$

The centre of gravity of the area A^* is defined by

$$X_G = \frac{\int X\, dA}{\int dA} = \frac{\int X\, dA}{A^*}; \qquad Y_G = \frac{\int Y\, dA}{A^*}$$

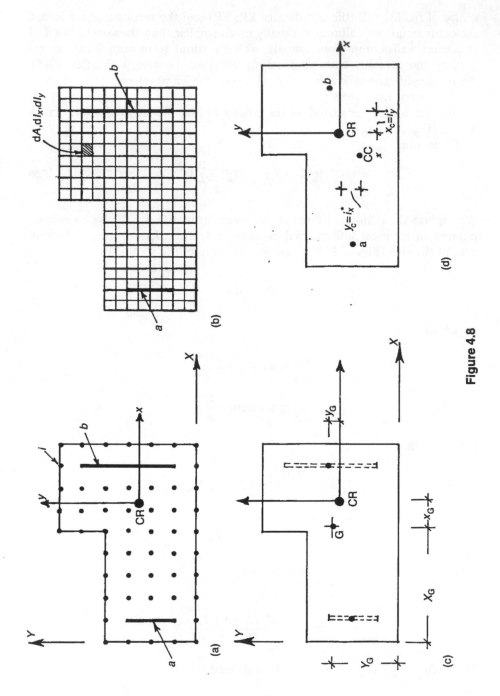

Figure 4.8

yielding (Figure 4.8c)

$$\left.\begin{aligned} XR &\cong \frac{I_x^{col} X_G + I_{x_a} X_a + I_{x_b} X_b}{I_x^{col} + I_{x_a} + I_{x_b}} \\ YR &\cong \frac{I_y^{col} Y_G + I_{y_a} Y_a + I_{y_b} Y_b}{I_y^{col} + I_{y_a} + I_{y_b}} \end{aligned}\right\} \tag{4.10}$$

We may interpret these formulae geometrically as follows: in order to compute the position of the centre of rigidity CR we can replace the columns by a single **resultant column** having the moments of inertia $I_x^{col} = \sum I_{x_i}$ and $I_y^{col} = \sum I_{y_i}$, positioned in the centre of gravity G of the area of the slab.

We recall that in most cases the centre of mass CM coincides with the centre of gravity G of the slab's area.

Similarly

$$\sum (I_{x_i} x_i^2) \cong \int x^2 \, dA \left(\frac{dI_x}{dA}\right) \cong I_x^{col} \int \frac{x^2 \, dA}{A^*}$$

$$\sum (I_{y_i} y_i^2) \cong \int y^2 \, dA \left(\frac{dI_y}{dA}\right) \cong I_y^{col} \int \frac{y^2 \, dA}{A^*}$$

but $\int x^2 \, dA$ and $\int y^2 \, dA$ represent the moments of inertia of the area A^* with respect to axes passing through the centre of rigidity CR:

$$\int x^2 \, dA = I_y^*; \qquad \int y^2 \, dA = I_x^*$$

This yields

$$\sum (I_{x_i} x_i^2) \cong \frac{I_x^{col} I_y^*}{A^*} = I_x^{col} i_y^{*2}$$

$$\sum (I_{y_i} y_i^2) \cong \frac{I_y^{col} I_x^*}{A^*} = I_y^{col} i_x^{*2}$$

The torsion constant C_0 becomes

$$C_0 = C_{col} + C_a + C_b$$

where

$$\left.\begin{aligned} C_{col} &= I_x^{col} i_y^{*2} + I_y^{col} i_x^{*2} = I_x^{col} x_c^2 + I_y^{col} y_c^2 \\ C_a &= I_{x_a} x_a^2 + I_{y_a} y_a^2 + \frac{Gh^2 I_{t_a}}{12\,E} \\ C_b &= I_{x_b} x_b^2 + I_{y_b} y_b^2 + \frac{Gh^2 I_{t_b}}{12\,E} \\ x_c &= i_y^* = \sqrt{\left(\frac{I_y^*}{A^*}\right)}; \qquad y_c = i_x^* = \sqrt{\left(\frac{I_x^*}{A^*}\right)} \end{aligned}\right\} \tag{4.11}$$

We may interpret these formulae geometrically as follows: in order to compute the torsion constant C_0 we can replace the columns by a single **resultant column** (I_x^{col}, I_y^{col}) positioned in a point CC (Figure 4.8d), at a distance defined by the coordinates

$$x_c = i_y^*; \qquad y_c = i_x^*$$

from the centre of rigidity CR.

We note that the sign of the coordinates x_c, y_c has no practical significance.

When the slab is divided into several zones with significant differences between the rigidities of the columns, we have to consider separately each zone (I, II) and compute the geometrical data accordingly:

$$
\left.
\begin{aligned}
XR &\cong \frac{I_{x_{(I)}}^{col} X_{G_{(I)}} + I_{x_{(II)}}^{col} X_{G_{(II)}} + I_{x_a} X_a + I_{x_b} X_b}{I_{x_{(I)}}^{col} + I_{x_{(II)}}^{col} + I_{x_a} + I_{x_b}} \\[2mm]
YR &\cong \frac{I_{y_{(I)}}^{col} Y_{G_{(I)}} + I_{y_{(II)}}^{col} Y_{G_{(II)}} + I_{y_a} Y_a + I_{y_b} Y_b}{I_{y_{(I)}}^{col} + I_{y_{(II)}}^{col} + I_{y_a} + I_{y_b}}
\end{aligned}
\right\}
\qquad (4.12)
$$

Note that when the centre of rigidity is close to the centre of mass, the analysis is very sensitive to inaccuracies occurring in the computation of the eccentricity, but then the whole problem of torsion is not important.

When structural walls or cores with high rigidities are present, we may compute the position of the centre of rigidity CR and the torsion constant C_0 by neglecting the presence of the columns:

$$
\left.
\begin{aligned}
XR &\cong \frac{I_{x_a} X_a + I_{x_b} X_b}{I_{x_a} + I_{x_b}}; \qquad YR \cong \frac{I_{y_a} Y_a + I_{y_b} Y_b}{I_{y_a} + I_{y_b}} \\[2mm]
C_0 &\cong C_a + C_b
\end{aligned}
\right\}
\qquad (4.13)
$$

□ **Numerical example 4.2**

Consider the slab shown in Figure 4.7. The coordinates of the centre of rigidity and the torsional shear forces will be computed according to the approximate method. As the columns in zones I and II are very different, we shall divide the area of the slab into two zones (Figure 4.9) and compute accordingly the coordinates XR, YR (equation (4.12)).

The areas of the slabs:

$$A_{(I)}^* = 300\ \text{m}^2; \qquad A_{(II)}^* = 300\ \text{m}^2$$

Centres of gravity:

$$X_{G_{(I)}} = 10{\cdot}0\ \text{m}; \qquad Y_{G_{(I)}} = 7{\cdot}5\ \text{m}$$

$$X_{G_{(II)}} = 27{\cdot}5\ \text{m}; \qquad Y_{G_{(II)}} = 10{\cdot}0\ \text{m};$$

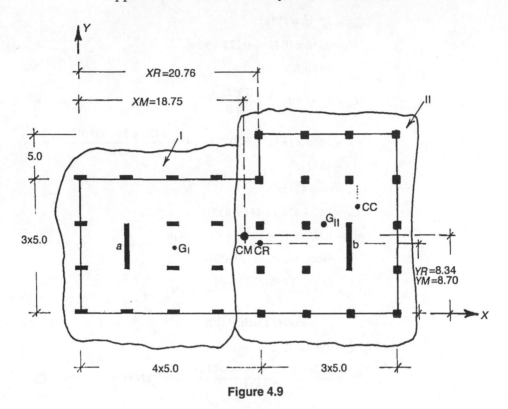

Figure 4.9

Total moments of inertia of columns:

$$I_{x_{(1)}}^{col} = 0.0252\,m^4; \qquad I_{y_{(1)}}^{col} = 0.1792\,m^4$$

$$I_{x_{(11)}}^{col} = 0.0938\,m^4; \qquad I_{y_{(11)}}^{col} = 0.0938\,m^4$$

The centre of rigidity CR:

$$X R = \frac{I_{x_{(1)}}^{col} X_{G_{(1)}} + I_{x_{(11)}}^{col} X_{G_{(11)}} + I_{x_a} X_a + I_{x_b} X_b}{I_x^{col} I_{x_a} I_{x_b}}$$

$$= \frac{0.0252 \times 10 + 0.0938 \times 27.5 + 1.856 \times (5 + 35)}{0.0252 + 0.0938 + 2 \times 1.856} = 20.12\,m$$

(accurate result: 20·08 m)

$$YR = \cdots = 8.34\,m \text{ (accurate result: 8·43 m)}$$

Area of the slab: $A^* = 600\,m^2$

The moments of inertia of the slab with respect to axes passing through CR:

$$I_x^* = 16562\,m^4; \qquad I_y^* = 61562\,m^4$$

$$x_c = i_y^* = 10.12; \qquad y_c = i_x^* = 5.26\,m$$

$$C_{\text{col}} = I_x^{\text{col}} x_c^2 + I_{y_c}^{\text{col}} y_c^2$$

$$= 0.119 \times 10.12^2 + 0.273 \times 5.25^2$$

$$= 19.7 \, \text{m}^6$$

$$C_a = I_{x_a} x_a^2 + I_{y_a} y_a^2 + \frac{G h^2 I_{t_a}}{12 E}$$

$$= 1.856 \times 15.12^2 + 0.0033 \times 0.84^2 + \frac{0.425 \times 3.0^2 \times 0.0133}{12}$$

$$= 424.3 \, \text{m}^6$$

$$C_b = \cdots = 411 \, \text{m}^6$$

$$C_0 = 19.7 + 424.3 + 411 = 855 \, \text{m}^6$$

$$M^T = 1000(20.12 - 18.75) = 1370 \, \text{kN m}$$

(accurate result: 1330 kNm).

$$\Delta V_{y_a}^T = \frac{1370 \times 1.856 (20.12 - 5)}{855} = 45 \, \text{kN}$$

(accurate result: 43 kN)

$$\Delta V_{x_a}^T = 0.005 \, \text{kN}$$

$$M_{t_a} = \frac{1370 \times 0.425 \times 3.0^2 \times 0.0133}{12 \times 855} = 0.007 \, \text{kN m} \qquad \square$$

4.3 A critical review of Lin's theory

4.3.1 EFFECT OF VERTICAL DEFORMABILITY OF SLABS

A basic assumption of Lin's theory implies the neglect of the vertical deformability of the slabs. This leads to two anomalies.

First, according to Lin's theory, torsional shear forces increase in proportion with the distance from the centre of rigidity. An accurate analysis displays a significant decrease of these forces close to the facades. In order to assess the order of magnitude of this effect, we shall refer to the structure shown in Figure 4.10a (12 storeys, slabs $60 \times 32 \times 0.20$ m, columns 0.40×0.40 m on a mesh of 5×4 m and a central core; shallow beams have been assumed in order to emphasize the effect of the vertical deformability of the slabs). The distribution of the torsional shear forces along the radii at each floor is displayed in Figure 4.10b. The maximum forces do not occur along the facades.

This phenomenon is due to the fact that the effective width of the slab strip acting together with the facade columns is much smaller than the corresponding strip acting together with the central columns; this leads to significant decreases in the rigidities of the facade columns.

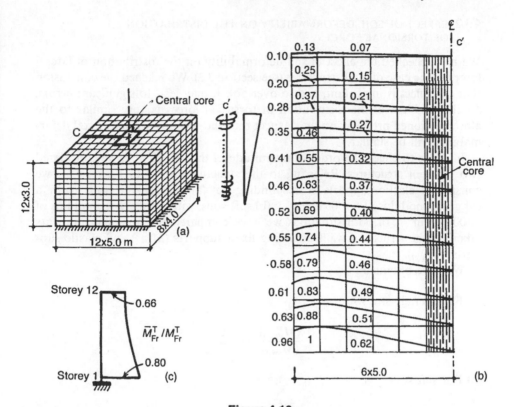

Figure 4.10

Mazilu (1989) noted that, in the 1977 and 1986 earthquakes in Romania, relatively small torsional forces acted on the corner columns of structures subjected to general torsion. This may have been due to the vertical deformability of the slabs and the subsequent low rigidity of the facade columns.

Second, the vertical deformations of the slabs lead to a decrease in the rigidity of the columns with respect to the rigidity assessed by Lin's theory; this means that the ratio of the moment of torsion taken by the columns is smaller than the ratio assumed by Lin's theory. By referring to the model shown in Figure 4.10, we obtain according to Lin's theory

$$M^T = M_{sw}^T \text{ (taken by the core)} + M_{Fr}^T \text{ (taken by the columns).}$$

An accurate analysis yields

$$M^T = \bar{M}_{sw}^T + \bar{M}_{Fr}^T$$

where $\bar{M}_{Fr}^T = (0.70\ldots0.80)\,M_{Fr}^T$. This effect increases along the height of the structure (Figure 4.10c).

4.3.2 EFFECT OF SOIL DEFORMABILITY ON THE DISTRIBUTION OF TORSIONAL FORCES

We have checked the effect of soil deformability on the distribution of lateral forces in the case of dual structures (see section 3.3). We reached the conclusion that this effect is important, and to overlook it may lead to significant errors. As the problem of the distribution of torsional forces is very similar to the aforementioned problem, we may expect that the effect of neglecting soil deformability will be similar.

In order to check the order of magnitude of this effect, we have considered two different structures. Referring to the structure shown in Figure 4.11, we considered elastic supports corresponding to: (a) average-stiff soils, with a subgrade modulus of $60\,000\,\mathrm{kN\,m^{-3}}$; (b) soft soils, with a subgrade modulus of $20\,000\,\mathrm{kN\,m^{-3}}$. In both situations we have compared the torsional moments taken by the structural walls assuming fixed supports $(M_{sw,f}^T)$ and elastic supports $(M_{sw,el}^T)$.

We have obtained:

(a) For average-stiff soils:

$$\frac{(M_{sw,el}^T)}{(M_{sw,f}^T)} = 0.53$$

(b) For soft soils:

$$\frac{(M_{sw,el}^T)}{(M_{sw,f}^T)} = 0.34$$

We see, in the considered structure, that the structural walls take only $\frac{1}{3}-\frac{1}{2}$ of the torsional forces yielded by a 'classical solution' (where we overlook soil deformability).

In Figure 4.12, the structure has two basement floors, surrounded by RC structural walls and eight storeys above the ground floor. An RC core $10 \times 5 \times 0.2\,\mathrm{m}$ is placed eccentrically. Columns $0.5 \times 0.5\,\mathrm{m}$ are positioned on a grid $5 \times 5\,\mathrm{m}$; the slabs are $0.2\,\mathrm{m}$ thick. A concentrated force normal to the longitudinal axis of symmetry acts in the centre of the roof slab.

We considered elastic supports corresponding to: (a) stiff soils $(k_s = 100\,000\,\mathrm{kN\,m^{-3}})$; (b) soft soils $(k_s = 20\,000\,\mathrm{kN\,m^{-3}})$. By comparing the torsional moments with those yielded by an analysis performed on the assumption of fixed ends we obtained:

(a) For stiff soils:

$$\frac{M_{sw,el}^T}{M_{sw,f}^T} = 0.64$$

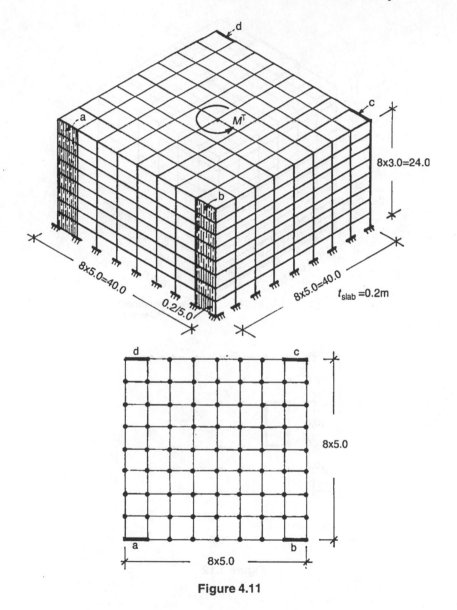

Figure 4.11

(b) For soft soils:

$$\frac{M^{\mathrm{T}}_{\mathrm{sw.el}}}{M^{\mathrm{T}}_{\mathrm{sw.f}}} = 0.38$$

We have to point out that in both examples we also considered given loads for fixed ends and elastic supports. Actually, the seismic forces are smaller

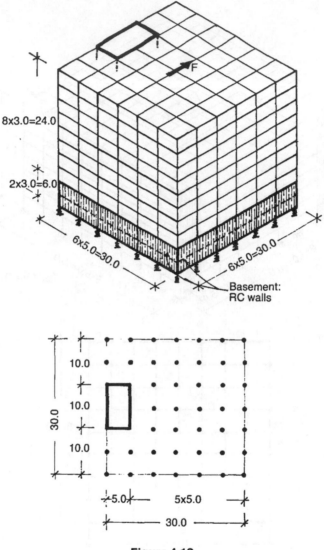

Figure 4.12

where elastic supports are taken into account; also, the bending moments will differ due to the change in the positions of the zero moment points.

In conclusion: considering the soil deformability leads to significantly smaller torsional forces than those yielded by 'classical analysis', based on the assumption of fixed ends. The decrease can be quantified by referring to the ratio e_{cl}/e, where e denotes the eccentricity of the resultant forces yielded by

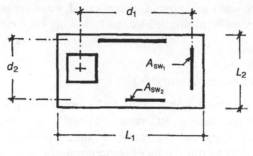

Figure 4.13

assuming fixed ends and e_{el} the eccentricity yielded by assuming a deformable soil.

By admitting conservatively $e_{el}/e = 0.6$ for stiff soils ($k_s = 100\,000\,\mathrm{kN\,m^{-3}}$) and $e_{el}/e = 0.4$ for soft soils ($k_s = 20\,000\,\mathrm{kN\,m^{-3}}$) and considering a linear variation of the ratio e_{el}/e between these limits, we obtain

$$\frac{e_{el}}{e} \cong 0.4 + \frac{0.2 \times (k_s - 20\,000)}{80\,000} \qquad (4.14)$$

where k_s is expressed in $\mathrm{kN\,m^{-3}}$.

In the case of foundations on piles we can consider (see Appendix C):

- for a small pile diameter ($D_p = 0.40\,\mathrm{m}$), an equivalent spread foundation on stiff and very stiff soils; by considering $k_s = 100\,000 - 200\,000\,\mathrm{kN\,m^{-3}}$, we obtain $e_{el}/e = 0.6 - 0.85$.
- for a large pile diameter ($D_p = 1.50\,\mathrm{m}$), an equivalent spread foundation on soft and regular soils; by considering $k_s = 20\,000 - 60\,000\,\mathrm{kN\,m^{-3}}$, we obtain $e_{el}/e = 0.40 - 0.50$.

4.4 First screening of existing dual structures subject to torsion

First screening of dual structures implies their classification from the point of view of their regularity in the horizontal plane (see Chapter 7).

At this end we define the **torsional index** (Figure 4.13):

$$\mathrm{TI} = \mathrm{S}\,\frac{\sum(A_{sw}\,d)}{L_{av}\sum A_{sw}} \qquad (4.15)$$

where A_{sw} denotes the area of an RC structural wall, belonging to a pair of parallel walls (where the walls are not identical we choose the wall with minimum area), d is the distance between the pair of walls, and L_{av} is the average horizontal dimension of the slab; in the case of irregular slabs $L_{av} = \mathrm{perimeter}/4$.

Where masonry walls are present, the areas A_{sw} are replaced by the areas A_m of the masonry walls, multiplied by the following correction factors:

- reinforced masonry: 0·6;
- infilled frames: 0·4;
- plain bricks or stone masonry: 0·3.

We have checked structures with 8, 10 and 12 storeys with columns 0.50×0.50, slabs 0.20 m thick and various RC structural walls with various deformable soils, and we have compared the ratios torsional shear forces/translational shear force. According to the results of these computations, we propose tentatively the following classification.

A **regular structure** is defined as either a symmetrical or a nearly symmetrical structure in both main directions, or as an asymmetrical one, but with a torsional index $TI > 2$ (when the pairs of parallel walls are positioned in one direction only, Figure 4.14a) or $TI > 1$ (when pairs of parallel walls are positioned in both main directions, Figure 4.14b). Note that in the case shown in Figure 4.14b, at least two parallel walls are subjected to torsional forces only, whereas in the case shown in Figure 4.14a the existing parallel walls are loaded simultaneously by translational and torsional forces (Paulay and Priestley, 1992).

A **moderately irregular structure** is an asymmetrical one with a torsional index $1 \leqslant TI \leqslant 2$ (Figure 4.14a) or $0.5 \leqslant TI \leqslant 1$ (Figure 4.14b).

All other structures are defined as **significantly irregular**.

Taking into account the results of comparative computations, we proposed an increase of the translational forces in the columns close to the perimeter, due to the presence of torsional forces (Scarlat, 1993) as follows:

15% for irregular structures;
25% for moderately irregular structures;
40% for significantly irregular structures.

As noted above, most modern structures require a minimum accidental eccentricity $e = 0.05 L$ to be taken into account for symmetrical structures

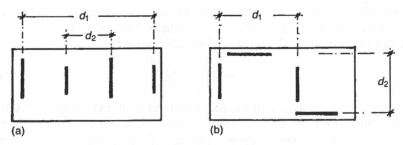

Figure 4.14

(L is the length of the building, normal to the given forces). The CEB-85 code proposes, as an alternative, multiplying the translational forces by the factor (Figure 4.15a)

$$\frac{1+0.6x}{L} \tag{4.16}$$

This leads to a maximum increase in the translational forces of 30%.

Accurate analyses performed by taking into account the soil deformability show that such an increase is equivalent to the presence of a quite strong eccentric core at a distance of about $L/4$ from the centre of masses (Figure 4.15b). This is obviously an exaggeration; we consider that the proposed increase in the translational forces of maximum 15% is more realistic.

4.5 Vertical grids composed of structural walls and slabs

In order to assess the stresses that develop in the slabs due to horizontal forces, we have to refer to a vertical grid formed by the structural walls and the slabs (Figure 4.16a,b). Obviously, we have to take into account the deformability of both types of elements: structural walls and slabs.

For an approximate analysis, we can refer to each slab separately, as beams on elastic supports (Figure 4.16c). The rigidity of the supports (k) is determined by loading each wall at the considered level by a unit horizontal force (Figure 4.16b). This procedure neglects the interaction of the structural walls.

When only two structural walls are present, the simplified procedure leads to statically determined structures; in this case the rigidity of the supports, as well as the effect of the shear forces, are not relevant (Figure 4.16d); we can consider the beams as resting on simple supports.

A procedure for assessing the distribution of the horizontal forces among structural walls by using an approach based on the use of difference equations has been proposed by Rutenberg and Dickman (1993).

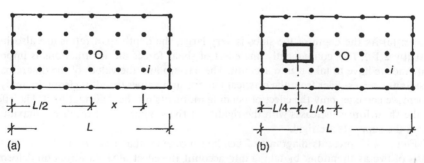

Figure 4.15

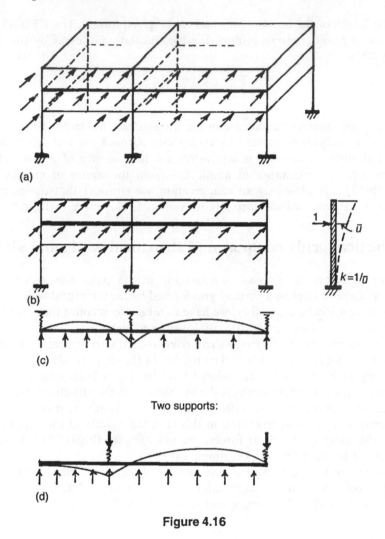

(a)

(b)

(c)

Two supports:

(d)

Figure 4.16

Notes: (a) As the depth of the slabs is very large, the depth/span ratios are also high (usually 2/1 ... 1/1); consequently, the effect of shear forces on deformations is important, and we have to take it into account. The pattern and the values of the diagrams of resultant stresses differ from the classical pattern and values (resulting from an analysis where we consider only the effect of bending moments on the deformations). The effect of the shear forces increases with the rigidity of the supports, and reaches a maximum when the supports are rigid.

Figure 4.17 presents diagrams of bending moments that have been drawn on the basis of two assumptions: by taking into account the effect of shear forces on deformations and by neglecting this effect. Rigid supports and elastic supports ($kl^3/EI = 20$ and $kl^3/EI = 2$) have been checked (I: moment of inertia of the beam). When elastic

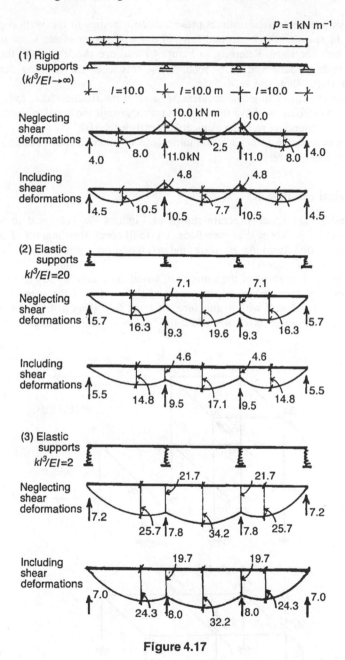

Figure 4.17

supports are considered, the negative moments decrease significantly, and usually only positive moments need to be taken into account. These are much greater than the moments resulting from an analysis where rigid supports have been assumed and reach, in the case of equal spans, values of $pl^2/5 \dots pl^2/3$.

(b) The resisting elements sometimes present discontinuities in the vertical plane. We then have to transfer the horizontal forces from one group of elements to another group through the slabs. Referring to Figure 4.18, above the level 3·00, the resisting elements are 2a3a–5a6a and 2b3b–5b6b. Below this level, the resisting elements are 1a2a–4a5a, 1b2b–4b5b and 1c2c–5b5c.

Obviously, the storey torsional moment M^T and its distribution have to be modified accordingly. Moreover, we have to ensure that we transmit the 'local moments' M' and M'' to the corresponding resisting elements beneath level 3.00.

As shown in section 6.3, such discontinuities lead to concentrations of seismic stresses and possible local distress.

☐ Numerical example 4.3

Let us consider the space structure shown in Figure 4.19a subjected to a uniform horizontal load $p = 1\,\text{kN}\,\text{m}^{-1}$ at each floor. We shall check the diagram M of the slab at level 6·00 (second floor). An accurate analysis, based on a grid analysis, leads to the diagram shown in Figure 4.19b.

An approximate analysis of the same slab, based on a continuous beam on elastic supports, must be preceded by an assessment of the rigidity of the supports. We then load each structural wall with a unit force at level 6·00 (Figure 4.19c) to obtain the deflections at the same level: $\bar{u} = 1·53 \times 10^{-6}\,\text{m}\,\text{kN}^{-1}$ (for the end supports) and

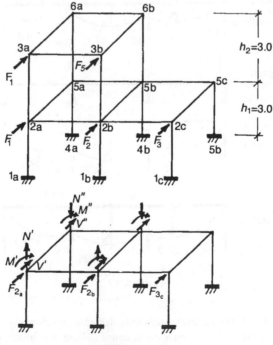

Figure 4.18

$\bar{u} = 1.02 \times 10^{-6}\,\mathrm{m\,kN^{-1}}$ (for the intermediate support), and the corresponding rigidities (spring constants) $k = 1/\bar{u} = 654\,000\,\mathrm{kN\,m^{-1}}$ and $980\,000\,\mathrm{kNm^{-1}}$ respectively. The resulting diagrams M are displayed in Figure 4.19d.   □

4.6 One-storey industrial buildings: bracing conditions

Consider a fragment of a one-storey industrial building, between a gable wall and an expansion joint (Figure 4.20a), on rigid supports; the structure is completely braced if no side sways are possible when rigid walls are assumed. In order to check the condition of complete bracing, we replace the structure by a space pin-jointed truss (Figure 4.20b); a minimum number of bars are considered for each wall, sufficient to ensure the bracing. The structure is completely braced if the pin-jointed truss satisfies the condition of geometrical fixity:

$$b + s \gtreqqless 3j \qquad (4.17)$$

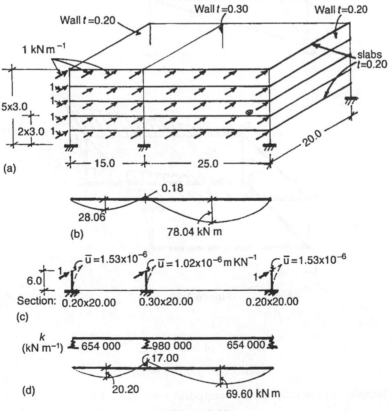

Figure 4.19

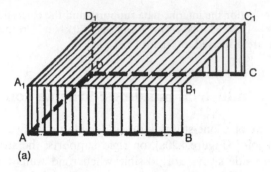

(a)

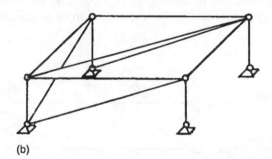

(b)

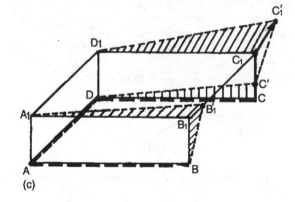

(c)

Figure 4.20

In our case:

$$\text{Number of bars } b = 12$$

$$\text{Number of joints } j = 8$$

$$\text{Number of restraints due to supports } s = 4 \times 3 = 12$$

$$12 + 12 = 3 \times 8$$

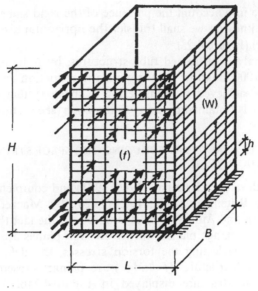

Figure 4.21

As a result, only axial forces have to be taken into account. When a significant settlement of the supports occurs, the bracing effect vanishes and bending moments develop along the edges (Figure 4.24c).

4.7 Framed tube structures: approximate assessment of stresses

4.7.1 INTRODUCTION

A new type of structure intended for very high-rise buildings (in excess of 30–40 stories) was developed in the 1970s, namely **framed tube structures**.

In this structural system, the horizontal forces are taken by a dense perimetral grid of columns and spandrels (Figure 4.21). The interior elements are designed to take only vertical forces. Columns with depths of 1–3 m are placed at 4–5 m distance. Spandrels 0.25–0.75 m thick and 1–1·5 m deep are placed at each floor.

If we replace the exterior grid by a continuous wall, a shell is obtained with a thickness-to-length ratio of $\frac{1}{50}-\frac{1}{100}$ (Taranath, 1988). Consequently, the structure behaves like a thin walled tube; a typical stress distribution is displayed in Figure 4.22, and is compared with a linear distribution yielded by the classical analysis, based on the assumptions of strength of materials.

A significant stress concentration due to the 'shear lag effect' occurs around the corners.

In order to take into account the presence of the rigid slabs without adding supplementary unknowns, we shall transfer the horizontal distributed loads to the side planes (*w*) (Figure 4.23).

An accurate analysis of framed tube structures, by finite elements, usually involves 5000–50 000 unknowns. A significant reduction in the number of unknowns can be achieved by considering a space system of bars; a high degree of accuracy is obtained by using **finite joints** (see section 2.3.2).

4.7.2 APPROXIMATE ANALYSIS BY REDUCING THE SPACE STRUCTURE TO A PLANAR STRUCTURE

The method was developed by Khan (1966, 1971) and completed by Coull and Subedi (1971), by Rutenberg (1972, 1973, 1974), by Mazzeo and De Fries (1972) and by Khan and Amin (1973). It is based on the fact that the displacements normal to the flange planes and to the web planes are negligible (Figure 4.24a). By also neglecting the torsion stresses, the deformations of the component plane substructures (*f* and *w*) are planar; subsequently, we may consider the planar structure displayed in Figure 4.24b. The connections

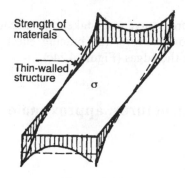

Figure 4.22

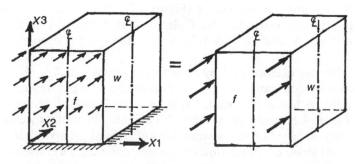

Figure 4.23

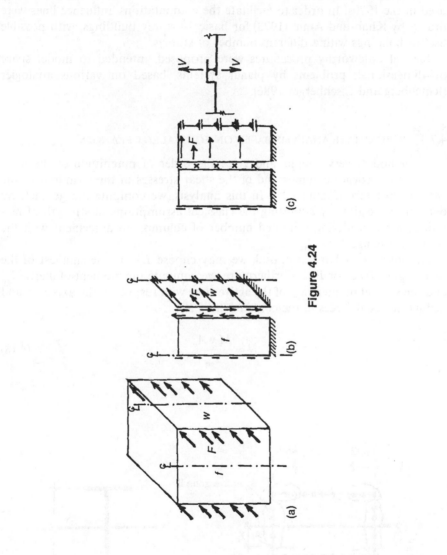

Figure 4.24

between the substructures f and w are meant to transfer vertical shear forces only (Figure 4.24c). These forces increase with the vertical rigidity of the flange substructure f: a rigid flange substructure prevents the free vertical displacements of the corner joints of the web substructure. The method was extensively used in the 1970s. In order to facilitate the computations, influence lines were drawn by Khan and Amin (1973) for basic 10-storey buildings, with possible use for buildings with a different number of storeys.

Several noteworthy procedures were proposed, intended to model space (tri-dimensional) problems by planar systems, based on various analogies (Rutenberg and Eisenberger 1986).

4.7.3 APPROXIMATE ANALYSIS BASED ON EQUIVALENT FLANGES

This method is very simple, and gives the order of magnitude of the axial forces in the corner columns and of the shear stresses in the spandrels of the web substructure (Figure 4.25). In this analysis, we compute the geometrical data of the section by admitting the classical assumptions of strength of materials, but considering a limited number of columns, in agreement with the equivalent flange $L_{eq} < L$.

According to Khan's proposal, we may choose L_{eq} as the smallest of the two lengths: $H/10$ or $B/2$. Furthermore, we compute the moment of inertia I_{eq} and the statical moment S_{eq} of the **active zone** with respect to the axis $x{-}x$ and deduct the axial forces in the columns (area A_c):

$$N = \frac{M\,y\,A_c}{I_{eq}} \tag{4.18}$$

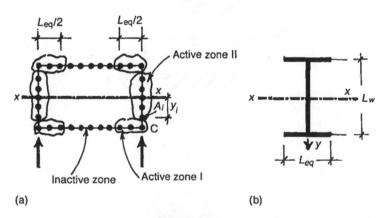

(a)   (b)

Figure 4.25

and the vertical shear forces V:

$$V_L = \frac{V S_{eq} h}{I_{eq}} \qquad (4.19)$$

where h denotes the storey height.

The proposed method gives satisfactory results for the maximum axial forces in the corner columns and for the maximum vertical shear forces in the bottom beams of the web substructure on the neutral axis $(x-x)$; however, because of the shear lag effect, higher vertical shear forces may occur close to the corners.

☐ **Numerical example 4.4**

The framed tube shown in Figure 4.26a is subjected to a lateral uniformly distributed load. Taking into account the in-plane rigidity of the slabs, we shall consider the lateral loads as linear loads distributed along the edges: let be $p = 4 \times 3$ kN m^{-1}. The total shear force results in $V_{max} = 1080$ kN, and the total overturning moment $M_{TOT} = 1080 \times 90/2 = 48\,600$ kN m. We compute the statical data for half of the given structure (a channel-type section). $V_{max} = 1080/2 = 540$ kN; $M_{max} = 48\,600/2 = 24\,300$ kN m.

The equivalent width of the flange (Figure 4.26b):

$$\frac{H}{10} = \frac{90}{10} = 9\text{ m}; \qquad \frac{B}{2} = \frac{24}{2} = 12\text{ m}; \qquad L_{eq} = 9\text{ m}$$

on each side, $L_{eq}/2 = 4.5$ m includes 2.5 columns. Let us denote the area of a column by A_c. The moment of inertia results in

$$I_{eq} = 2 \times 2.5 \times A_c \times 12^2 + 4 A_c (4^2 + 8^2) = 1040\, A_c$$

with the corresponding modulus of resistance

$$W_{eq} = \frac{1040\, A_c}{12} = 86.7\, A_c$$

The static moment with respect to the axis $x-x$:

$$S_{eq} = 2.5\, A_c \times 12 + A_c(4 + 8) + 0.5\, A_c \times 12 = 48\, A_c$$

The maximum axial force (at corners):

$$N = \frac{M_{max}\, A_c}{W_{eq}} = \frac{24\,300}{86.7} = 280\text{ kN (accurate analysis: } N = 304\text{ kN).}$$

The maximum vertical shear force (in the spandrels of the web substructure):

$$V_{L_{max}} = \frac{V_{max}\, h}{z}$$

$$z = \frac{I_{eq}}{S_{eq}} = \frac{1040}{48} = 21.66\text{ m}; \qquad h = 3\text{ m}$$

$$V_{L_{max}} = \frac{540 \times 3}{21.66} = 75\text{ kN}$$

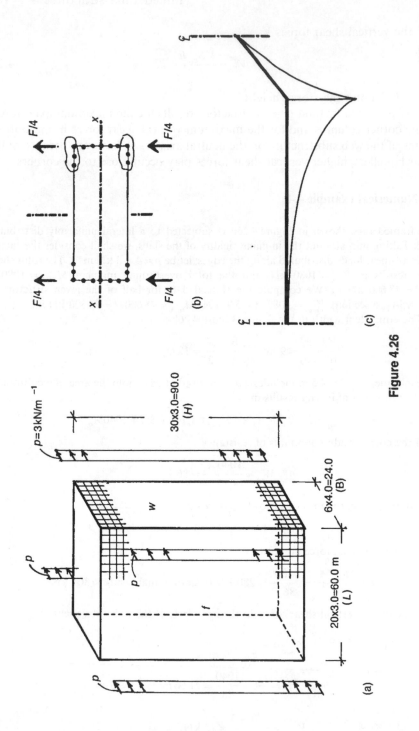

Figure 4.26

Accurate analysis (tri-dimensional finite element analysis):

$$V_{L_{max}} = 46 \text{ kN (in the neutral axis)} \dots 75 \text{ kN (close to the corner)}$$

The distribution of the vertical reactions on half of the wall yielded by the accurate analysis is displayed in Figure 4.26c; a significant shear lag effect is visible.   □

4.8 Axi-symmetric system of bars: evaluation of maximum stresses and deflections

4.8.1 INTRODUCTION

Consider a water tank (Figure 4.27) supported on a space, axi-symmetric frame (a regular polygon with n sides), subject to horizontal forces (either seismic forces or wind pressure). The resultant force F_T acts at height H_T; it may be replaced by the same resultant force F_T acting above the columns plus a resultant moment $M_k = F_T H_k/2$. As the bending moments M, the shear forces V and the horizontal deflections u due to the resultant moment M_k are usually negligible we may, when computing M, V and u, relate only to the effect of the resultant force F_T. The columns of the space frame are usually placed on a

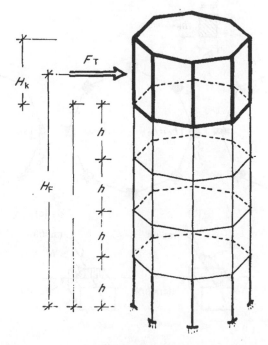

Figure 4.27

hexagon, an octagon or a dodecagon. They may be radial (Figure 4.28a), square (Figure 4.28b) or tangential (Figure 4.28c).

In the following, we shall give approximate values of the fixed end bending moments, at the base of the columns (radial moments M^r and tangential moments M^t, Figure 4.28d), maximum deflections and axial forces due to the resultant force F_T.

We note that in most cases the fixed end moments also represent the maximum moments

We assume that:

- the polygon beams are positioned at equal heights $H = m\,h$;
- the polygons maintain their initial form in the deflected shape (it has been demonstrated that this assumption is theoretically valid for the considered type of loading: Mutafolo, 1959);
- the columns are fixed at their base and at their top ends, in the tank (the structure of the tank is considered as a rigid body).

4.8.2 FIXED END MOMENTS

The fixed end moments depend on the given resultant force F_T, the type of polygon (number of sides n), the direction of the columns (radial, square or tangential), and the relative rigidities of the columns and beams.

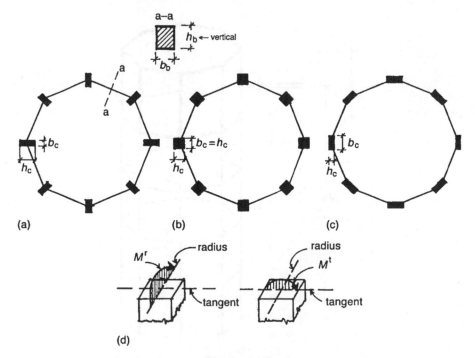

Figure 4.28

We have computed the fixed end moments M_f for three types of structure: hexagonal, octagonal and dodecagonal. For each type, we considered (Figure 4.28): radial columns ($h_c = 5b_c$), square columns ($h_c = b_c$) and tangential columns ($h_c = 0.2 b_c$). Beams of various rigidities were taken into account ('flexible beams', 0.20×0.20 m; 'average beams', 0.32×0.32 m; and 'stiff beams', 0.50×0.50 m). Space frames ranging from three to five storeys were considered.

The results are given in the form of graphs: $M_f^r(n/F_T h)$ and $M_f^t(n/F_T h)$ as functions of the relative rigidities k_b/k_c, where

$$k_b = \frac{I_b}{l}; \qquad k_c = \frac{I_c}{h_c}; \qquad I_b = \frac{b_b h_b^3}{12}; \qquad I_c = \frac{b_c h_c^3}{12} \qquad \text{(see Figure 4.28)}.$$

The graphs are shown in Figure 4.29.

After computing the maximum moments $M_f^r = M_{f_1}^r$ and $M_f^t = M_{f_1}^t$ we can obtain the corresponding moments in the other columns i according to the expressions (Figure 4.30)

$$M_{f_i}^r = M_{f_1}^r \cos(2\pi i/n); \qquad M_{f_i}^t = M_{f_1}^t \sin(2\pi i/n) \qquad (4.20)$$

4.8.3 MAXIMUM AXIAL FORCES

The axial forces are due to the effect of vertical loads and the effect of horizontal forces. In the following we shall relate only to these latter forces. The maximum axial forces are equal to the vertical reactions R_v: $N_{max} = R_v$.

We shall assume that the reactions vary linearly, in proportion to the distance d_i to the axis of symmetry S, normal to the resultant force F_T.

Equilibrium by moments about axis S (Figure 4.31) yields

$$M^{ext} = M^{R_v} + M^M$$

where $M^{ext} = F^T H_F$; M^{R_v} is the moment given by the vertical reactions R_v; M^M is the moment given by the fixed end moments acting on the columns, projected on the axis S; $d_i = r \sin \alpha_i = r \sin 2\pi i/n$; and $M^{R_v} = \sum R_{v_i} d_i$.

Examination of numerical examples indicate that we may accept

$$M^M \cong 0.1 \, M^{ext}$$

Accordingly:

$$F^T H_F = \frac{R_{v_{max}} \, r \, n}{2} + 0.1 \, F^T H_F \qquad (4.21)$$

$$N_{max} = R_{v_{max}} \cong \frac{1.8 \, F^T H_F}{n r}$$

In the remainder columns:

$$N_i = N_{max} \cos 2\pi \, i/n \qquad (4.22)$$

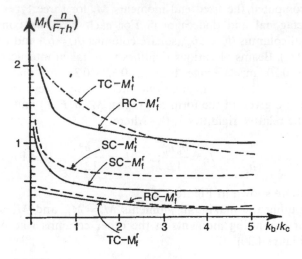

(a) Hexagon

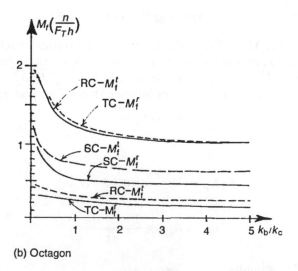

(b) Octagon

Figure 4.29

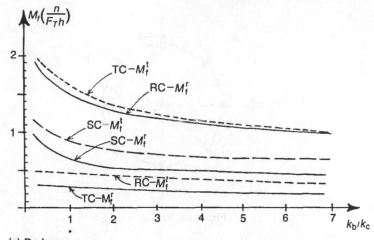

(c) Dodecagon
RC- Radial columns, SC- Square columns, TC- Tangential columns,

M^r - Radial moments, M^t- Tangential moments

Figure 4.29 Contd.

4.8.4 MAXIMUM HORIZONTAL DEFLECTIONS

We shall first compute the maximum horizontal deflections by assuming that the beams are rigid (u°_{max}). We shall then correct the deflections u°_{max} by taking into account the deformability of the beams.

Referring to Figure 4.32 and considering only one storey and assuming rigid beams (sliding fixed ends), we find:

$$\left.\begin{aligned}
F_T &= \sum_{1}^{n}(V^r_i \cos\alpha_i + V^t_i \sin\alpha_i) \\[2mm]
&= 12\,E\,u^{\circ}_{max}\left(\frac{I_t\sum_{1}^{n}\cos^2\alpha_i + I_r\sum_{1}^{n}\sin^2\alpha_i}{h^3}\right) \\[2mm]
&= \frac{12\,E\,u^{\circ}_{max}(I_t\,n/2 + I_r\,n/2)}{h^3} \\[2mm]
&= \frac{12\,E\,I_s\,u^{\circ}_{max}}{h^3}
\end{aligned}\right\} \tag{4.23}$$

where

$$I_s = \frac{n(I_r + I_t)}{2}; \qquad I_r = \frac{b_c\,h_c^3}{12}; \qquad I_t = \frac{h_c\,b_c^3}{12}$$

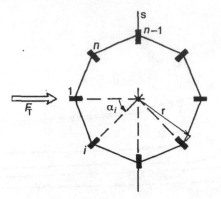

Figure 4.30

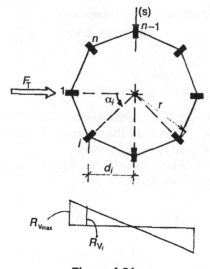

Figure 4.31

By considering m storeys ($H = m\,h$):

$$u^{\circ}_{max} = \frac{F_T\,m\,h^3}{12\,E\,I_s} \qquad (4.24)$$

We take into account the deformability of the beams by multiplying the deflection u°_{max} by the factor μ:

$$u_{max} = \mu\,u^{\circ}_{max}$$

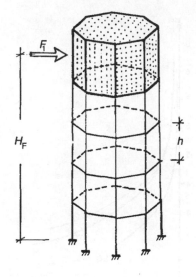

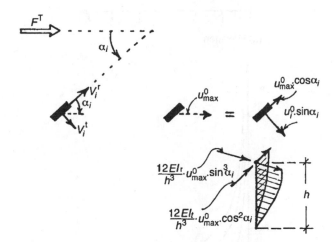

Figure 4.32

Graphs of the factor μ are shown in Figure 4.33, for the same cases that we considered previously.

Evaluation of maximum horizontal deflection also enables us to assess the fundamental period of free vibrations (T). Let

$$u_W = \frac{\mu\, W\, m\, h^3}{12\, E\, I_s} \tag{4.25}$$

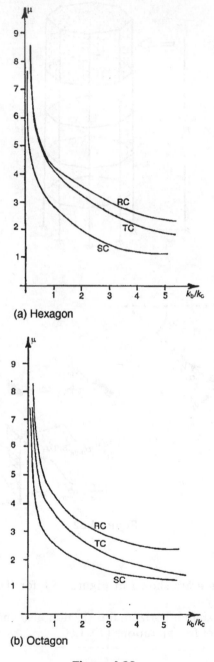

(a) Hexagon

(b) Octagon

Figure 4.33

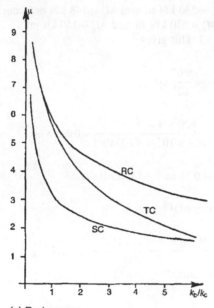

(c) Dodecagon
RC- Radial columns; SC- Square columns; TC- Tangential columns

Figure 4.33 Contd.

be the maximum horizontal deflection due to a horizontal force equal to the total weight of the tank (W), including its content. According to Geiger's formula (see Appendix A), $T = 2\sqrt{u_W}$, where u_W is given in m.

□ **Numerical example 4.5**

Consider the octagonal water tower shown in Figure 4.27 with the following data: radial columns 0.20×0.50 m, beams 0.32×0.32 m (average beams), radius $r = 4.0$ m, $H = 4h = 4 \times 4.0 = 16$ m, $H_F = 18$ m; $F_T = 800$ kN; $E = 3 \times 10^7$ kN m^{-2}.

$$k_b = \frac{I_b}{l} = \frac{0.32^4}{12 \times 3.06} = \frac{2.85}{10^4} m^3$$

$$k_c = \frac{I_c}{h} = \frac{0.20 \times 0.50^3}{12 \times 4.0} = \frac{5.21}{10^4} m^3$$

$$\frac{k_b}{k_c} = 0.55.$$

From Figure 4.29b, we read $M_f^r n/(F^T h) \cong 1.35$ and $M_f^t n/(F_T h) \cong 0.37$; $n/(F_T h) =$

$8/(800 \times 4) = 2 \cdot 5/10^3$. This gives $M_f^r = 540$ kN m and $M_f^t = 148$ kN m (accurate results, yielded by a space frame analysis: $M_f^r = 530$ kN m and $M_f^t = 150$ kn m).

From Figure 4.33b we read $\mu \cong 5 \cdot 3$. This gives

$$I_s = \frac{n(I_r + I_t)}{2} = \cdots = \frac{9 \cdot 67}{10^3} \, m^4$$

$$u_{max}^\circ = \frac{F_T m h^3}{(12 E I_s)} = \frac{800 \times 4 \times 4 \cdot 0^3}{(12 \times 3 \times 10^7 \times 9 \cdot 67/10^3)} = 0 \cdot 0589 \, m$$

$$u_{max} = 5 \cdot 3 \times 0 \cdot 0589 = 0 \cdot 31 \, m \qquad \text{(accurate result: } u_{max} = 0 \cdot 28 \, m\text{)}$$

The maximum axial force due to horizontal forces is

$$N_{max} \cong \frac{1 \cdot 8 \times F_T H}{n r} = \frac{1 \cdot 8 \times 800 \times 18 \cdot 0}{8 \times 4 \cdot 0} = 810 \, kN$$
$$\text{(accurate result:} N_{max} = 632 kN\text{)}.$$

Fundamental period:

$$W = 10\,000 \, kN; \qquad u_w = \frac{10\,000 \times 0 \cdot 31}{800} = 3 \cdot 875 \, m$$

$$T = 2\sqrt{3 \cdot 875} = 3 \cdot 94 \, s \quad \text{(accurate computation: } T = 3 \cdot 785 \, s\text{)}$$

In the case of tangential columns ($0 \cdot 50 \times 0 \cdot 20$ m) we obtain

$$\frac{k_b}{k_c} = 3 \cdot 43$$

$$M_f^r = 80 \, kN \, m \quad \text{(accurate: 75 kN m)}$$

$$M_f^t = 420 \, kN \, m \quad \text{(accurate: 454 kN m)}$$

$$u_{max} = \mu u_{max}^\circ = 2 \cdot 2 \times 0 \cdot 0589 = 0 \cdot 13 \, m \quad \text{(accurate: } 0 \cdot 16 \, m\text{)}$$

$$N_{max} = 810 \, kN \text{ (accurate: 675 kN)}$$

$$T = 2 \cdot 55 \, s \quad \text{(accurate: } 2 \cdot 82 \, s\text{)}$$

We note that the tangential orientation of the columns is preferable, in terms of stiffness, to the radial orientation of columns; it also provides a higher torsional stiffness.   □

Bibliography

Coull, A. and Subedi, N. (1971) Framed-tube structures for high-rise buildings. *Journal of Structural Division ASCE*, **97**, 2097–2105.

Jiang, W., Hutchinson, G. and Chandler, A. (1993) Definitions of static eccentricity for design of asymmetric shear buildings, *Engineering Structures*, **15** (3), 167–178.

Khan, F. (1966) Current trends in concrete high rise buildings, in *Proceedings Symposium on Tall Buildings*, University of Southampton, April, pp. 571–590.

Khan, F. (1971) Tendances actuelles dans la construction des immeubles de grande hauteur à structure en béton armé et en acier. *Annales Institut technique du bâtiment et des trauvax publics*, Suppl. au No. 281 (Mai), pp. 37–53.

Khan, F. and Amin, N. (1973) Analysis and design of framed tube structures for tall concrete buildings. *Journal of the American Concrete Institute*, **53** (SP36), 85–90.

Khan, F. and Stafford Smith, B. (1975) Restraining action of bracing in thin-walled open section beams. *Proceedings of the Institution of Civil Engineers. Part 2.* **59** (March), 67–78.

Lin, T. Y. (1951) Lateral forces distribution in a concrete building story. *Journal of the American Concrete Institute*, **23** (4), 281–294.

Mazilu, P. (1989) Behaviour of buildings during the 1977 and 1986 earthquakes in Romania, lecture at the Faculty of Civil Engineering, Technion, Haifa, November.

Mazzeo, A. and De Fries, A. (1972) Perimetral tube for 37-story steel building. *Journal of the Structure Division ASCE*, **98** (ST6), 1255–1273.

Mutafolo, M. (1959) Contribution to the analysis of multi-storey, axi-symmetric towers. Thesis for PhD, Institute of Civil Engineering, Bucarest (in Romanian).

Paulay, T. and Priestley, M. (1992) *Seismic Design of RC and Masonry Buildings*, J. Wiley, New York.

Rutenberg, A. (1972) Discussion of paper by A. Coull, N. Subedi (Framed-tube structures for high-rise buildings), *Journal of the Structural Division Proceedings of the ASCE*, **98** (ST4) 942–943.

Rutenberg, A. (1972) Discussion of paper by A. Mazzeo, A. De Fries (Perimetral tube for 37-story steel building), *Journal of the Structural Division Proceedings of the ASCE*, **99** (ST3) 586–588.

Rutenberg, A. (1974) Analysis of tube structures using plane programs, in *Proceedings of the Regional Conference on Tall Buildings*, Bangkok. pp. 397–413.

Rutenberg, A. (1980) Laterally loaded flexible diaphragm buildings. *Journal of the Structural Division Proceedings of the ASCE*, **106** (ST9), 1969–1973.

Rutenberg, A. and Eisenberger, M. (1986) Simple planar modeling of asymmetric shear buildings for lateral forces. *Computers and Structures*, **24** (6), 885–891.

Rutenberg, A. and Dickman, Y. (1993) Lateral load response of setback shear wall buildings. *Engineering Structures*, **15** (1), 47–54.

Scarlat, A. (1986) Approximate analysis of multistorey buildings, in 1er *Colloque National de Génie Parasismique*, St Rémy, pp. 6.1–6.10.

Scarlat, A. (1993) Soil deformability effect on rigidity-related aspects of multistorey buildings analysis. *Structure Journal of the American Concrete Institute*, **90** (2), 156–162.

Taranath, B. (1988) *Structural Analysis and Design of Tall Buildings*, McGraw-Hill, New York.

5 Pile foundations and retaining walls

5.1 The 'equivalent pile length' concept

A quick approximate assessment of the maximum moment and the maximum deflection of a pile subjected to a horizontal force can be performed by applying the **equivalent pile length** concept. In the following, we shall use the results obtained by Kocsis (1968) in the form adopted by the Ministry of Works and Development of New Zealand (1981) (cited by Dowrick, 1987). In order to characterize the mechanical properties of soils we may classify them as **cohesive** (e.g. clays) or **cohesionless** (e.g. sands).

For cohesive soils, we defined the horizontal soil deformability by using the **modulus of subgrade reaction** k_s, i.e. the normal stress corresponding to a unit horizontal displacement (see Appendix C). We may assume a constant value of k_s along the height of the pile (Figure 5.1). k_s is usually in the range $5000-70\,000\,\mathrm{kN\,m^{-3}}$. We define the **stiffness radius**

$$R_c = \left(\frac{E_p I_p}{k}\right)^{1/4} \quad \text{(m)} \tag{5.1}$$

where

$$k = k_s D_p \quad (\mathrm{kN\,m^{-2}}) \tag{5.2}$$

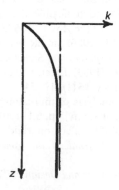

Figure 5.1

D_p is the diameter of the pile (m), I_p is the moment of inertia of the pile (m^4), and E_p is the modulus of elasticity of the pile (kN m^{-2}).

For cohesionless soils we assume that soil stiffness increases linearly with depth z (Figure 5.2); we characterize it by using the **unit subgrade reaction** n_s, defined as

$$n_s = \frac{k_s D_p}{z} \quad (\text{kN m}^{-3}) \tag{5.3}$$

n_s is usually in the range 1000–20 000 kN m^{-3}.

The stiffness radius is

$$R_n = \left(\frac{E_p I_p}{n_s}\right)^{1/5} \quad (\text{m}) \tag{5.4}$$

In the following, the piles are assumed flexible, i.e.

$$\frac{L}{R_c} \geqslant 4; \qquad \frac{L}{R_n} \geqslant 4 \tag{5.5}$$

where L denotes the length of the pile below the ground surface. We note that the modulus of subgrade for groups of piles may significantly decrease with the pile spacing (Davisson and Salley, 1970; Poulos, 1979).

We define the "equivalent pile lengths" in the following by referring to two situations, free-headed piles and fixed-headed piles.

For free-headed piles (Figure 5.3), the equivalent length L_m for the evaluation of the maximum moment is given by

$$M_{max} = F(L_m + a) \tag{5.6}$$

The equivalent length L_d for the computations of the maximum deflections is

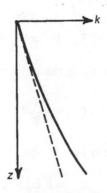

Figure 5.2

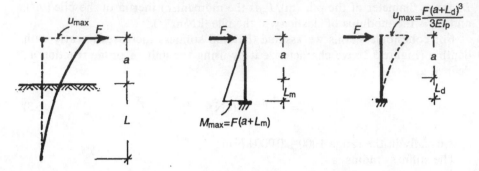

Figure 5.3

given by

$$u_{max} = \frac{F(L_d + a)^3}{3 E_p I_p} \tag{5.7}$$

L_m and L_d are determined as follows.

For cohesive soils:

$$\left. \begin{aligned} L_m &= 0 \cdot 5 \, R_c \\ L_d &= 1 \cdot 4 \, R_c \quad \text{for } a/R_c \geqslant 2 \\ & 1 \cdot 6 \, R_c \quad \text{for } a/R_c < 2 \end{aligned} \right\} \tag{5.8}$$

For cohesionless soils:

$$\left. \begin{aligned} L_m &= 0 \cdot 8 \, R_n \\ L_d &= 1 \cdot 8 \, R_n \quad \text{for } a/R_n \geqslant 1 \\ & 2 \cdot 2 \, R_n \quad \text{for } a/R_n < 1 \end{aligned} \right\} \tag{5.9}$$

For fixed-headed piles (Figure 5.4), the equivalent length L_m is given by

$$M_{max} = \frac{F \, L_m}{2} \tag{5.10}$$

The equivalent length L_d is given by

$$u_{max} = \frac{F \, L_d^3}{12 \, E_p \, I_p} \tag{5.11}$$

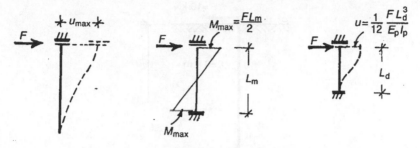

Figure 5.4

L_m and L_d are determined as follows.

For cohesive soils:

$$L_m = 1 \cdot 5\,R_c; . \quad L_d = 2 \cdot 2\,R_c$$

For cohesionless soils: (5.12)

$$L_m = 2 \cdot 0\,R_n; \quad L_d = 2 \cdot 5\,R_n$$

The radii R_c and R_n were defined by equations (5.1) and (5.4).

□ Numerical example 5.1

A free-headed pile with a diameter $D_p = 0 \cdot 80$ m and a length $L = 12 \cdot 5$ m is acted upon by a horizontal force $F = 10$ kN (Figure 5.5).

$$E_p\,I_p = 3 \times 10^7 \times 0 \cdot 0201 = 603\,000 \text{ kN m}^2.$$

(a) Cohesive soil

$$k_s = 10\,000 \text{ kN m}^{-3}; \; k = k_s\,D_p = 8000 \text{ kN m}^{-2} \quad R_c = \left(\frac{603\,000}{8000}\right)^{1/4} = 2 \cdot 94 \text{ m},$$

$$\frac{L}{R_c} = \frac{12}{2 \cdot 94} = 4 \cdot 25 > 4; \; a = 0; \; L_m = 0 \cdot 5 \times 2 \cdot 94 = 1 \cdot 47 \text{ m}; \; M_{max} = 10 \times 1 \cdot 47 = 14 \cdot 70 \text{ kN m}.$$

(An 'accurate analysis', performed by considering discrete horizontal springs, yields $M_{max} = 13$ kN m.)

$$\frac{a}{R_c} = 0 < 2$$

$$L_d = 1 \cdot 6 \times 2 \cdot 94 = 4 \cdot 70 \text{ m}$$

$$u_{max} = \frac{10 \times 4 \cdot 70^3}{3 \times 603\,000} = 0 \cdot 000\,57 \text{ m} = 0 \cdot 57 \text{ mm (accurate: } 0 \cdot 60 \text{ mm).}$$

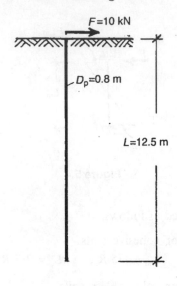

Figure 5.5

(b) Cohesionless soil

$$n_s = 2000\,\mathrm{kN\,m}^{-3}$$

$$R_n = (603\,000/2000)^{1/5} = 3\cdot13\,\mathrm{m}$$

$$\frac{L}{R_n} = 12\cdot5/3\cdot13 = 3\cdot99$$

As L/R is very close to 4 we shall relate to the pile as to a flexible one.

$$L_m = 0\cdot8 \times 3\cdot13 = 2\cdot50\,\mathrm{m}$$

$$M_{max} = 10 \times 2\cdot50 = 25\,\mathrm{kN\,m}\;(\text{accurate}: 24\cdot1\,\mathrm{kN\,m}).$$

$$\frac{a}{R_n} = 0 < 1; \qquad L_d = 2\cdot2 \times 3\cdot13 = 6\cdot89\,\mathrm{m}$$

$$u_{max} = \frac{10 \times 6\cdot89^3}{3 \times 603\,000} = 0\cdot0018\,\mathrm{m} = 1\cdot8\,\mathrm{mm}\;(\text{accurate}: 1\cdot6\,\mathrm{mm}). \qquad \square$$

☐ **Numerical example 5.2**

A fixed-headed pile with a diameter of $0\cdot40\,\mathrm{m}$ and a length $L = 10\,\mathrm{m}$ is acted upon by a horizontal force $F = 10\,\mathrm{kN}$ (Figure 5.6).

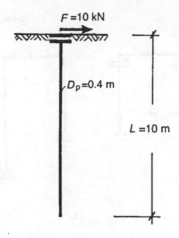

Figure 5.6

(a) Cohesive soil

$$k_s = 10\,000\,\text{kN m}^{-3}; \quad k = k_s D_p = 4000\,\text{kN m}^{-2}; \quad E_p I_p = 37\,700\,\text{kN m}^2$$

$$R_c = \left(\frac{37\,700}{4000}\right)^{1/4} = 1\cdot75\,\text{m} \qquad \frac{L}{R_c} = \frac{10}{1\cdot75} = 5\cdot71 > 4$$

$$L_m = 1\cdot5 \times 1\cdot75 = 2\cdot63\,\text{m}$$

$$M_{max} = \frac{10 \times 2\cdot63}{2} = 13\cdot15\,\text{kN m} \quad \text{(accurate: 12·10 kN m)}$$

$$L_d = 2\cdot2 \times 1\cdot75 = 3\cdot85\,\text{m}$$

$$u_{max} = \frac{10 \times 3\cdot85^3}{12 \times 37\,700} = 0\cdot0013\,\text{m} = 1\cdot3\,\text{mm} \quad \text{(accurate: 1 mm)}.$$

(b) Cohesionless soil

$$n_s = 2000\,\text{kN m}^{-3}$$

$$R_n = \left(\frac{37\,700}{2000}\right)^{1/5} = 1\cdot80\,\text{m}$$

$$L_m = 2\cdot0\,R = 2\cdot0 \times 1\cdot80 = 3\cdot60\,\text{m}$$

$$M_{max} = \frac{10 \times 3\cdot60}{2} = 18\,\text{kN m} \quad \text{(accurate: 20 kN m)}$$

$$L_d = 2\cdot5\,R = 4\cdot50\,\text{m}$$

$$u_{max} = \frac{10 \times 4\cdot50^3}{12 \times 37\,700} = 0\cdot002\,\text{m} = 2\,\text{mm} \quad \text{(accurate: 2·5 mm)} \qquad \square$$

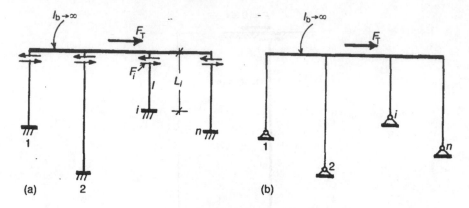

Figure 5.7

5.2 Distribution of lateral forces in a pile group

We often have to deal with the problem of distributing a given horizontal force among piles with different diameters and different lengths.

In the case of a group of columns with various sections and lengths with fixed base and top sliding fixed end, as in the case of rigid connecting beams (Figure 5.7a), subjected to a given horizontal force F_T, the solution is simple: the force F_T will be divided among the columns in proportion to their rigidities:

$$F_i = F_T \frac{I_i/L_i}{\sum I_i/L_i} \tag{5.13}$$

The same formula governs the distribution of the force F_T among columns when they are pinned at their lower ends (Figure 5.7b). In the case of elastic connecting beams equation (5.13) gives an approximate solution.

However, in the case of piles, the presence of soil resistance to horizontal displacements completely changes the picture (Figure 5.8): the distribution of the force F among piles according to their rigidities would be very far from the reality. Hence, in order to determine an acceptable distribution of the given horizontal force F_T among the piles, we shall refer to two cases, as follows.

The first case is a pile group with constant length in an identical soil, but with different diameters. For this purpose we consider a group of five piles, 10 m long, and diameters varying from 0·40 to 1·20 m (Figure 5.8). The piles are connected by a horizontal beam. The following alternatives are considered:

- a soft soil ($k_s = 20\,000\,\mathrm{kN\,m^{-3}}$) and a stiff soil ($k_s = 100\,000\,\mathrm{kN\,m^{-3}}$);
- a very stiff connecting beam (sliding fixed top ends) and a very flexible connecting beam (pinned top ends);

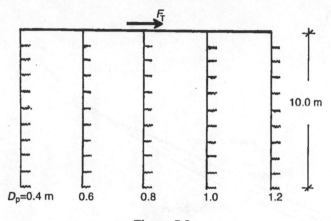

10.0 m

D_p=0.4 m   0.6   0.8   1.0   1.2

Figure 5.8

- soil resistance to horizontal displacements at top is neglected for the first 1 m and for the first 2 m.

The results are displayed graphically in Figure 5.9, as average curves F_i/F_T, where F_i is the horizontal force taken by the pile i, versus the diameters' ratio $D_{p_i}/D_{p_{max}}$, for soft and stiff soils.

We can assume the average curve to be a straight line (see Figure 5.10);

$$\frac{F_i}{F_T} \simeq \frac{D_{p_i}}{2\,D_{p_{max}}} - \frac{1}{10} \tag{5.14}$$

The second case is a group of micropiles with identical diameters but of different lengths.

We refer to a rather usual case, when a backfill (compacted or not) is placed on a sloping hard soil (often a rock-type soil). The micropiles will have different lengths, according to the slope of the hard soil.

In order to determine the distribution of the horizontal force F_T among the micropiles we consider (Figure 5.11) a group of five micropiles with a diameter of 0·35 m, placed on a backfill of 2·5–4·5 m and penetrating into the hard soil at depths of 1·5–3·5 m.

We consider the following alternatives:

- a very soft backfill ($k_s = 5000\,\mathrm{kN\,m^{-3}}$) on a very hard soil ($k_s = 200000\,\mathrm{kN\,m^{-3}}$); an average backfill ($k_s = 30000\,\mathrm{kN\,m^{-3}}$) on a hard soil ($k_s = 100000\,\mathrm{kN\,m^{-3}}$);
- a very stiff connecting beam (sliding fixed top ends) and a very flexible connecting beam (pinned top ends);
- the soil resistance to horizontal displacements at top is neglected for the first 1 m and for the first 2 m.

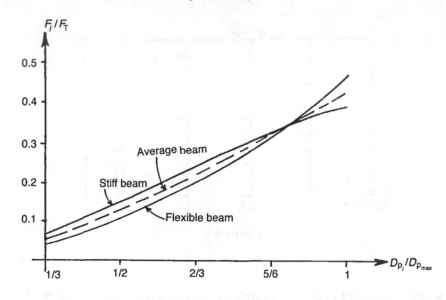

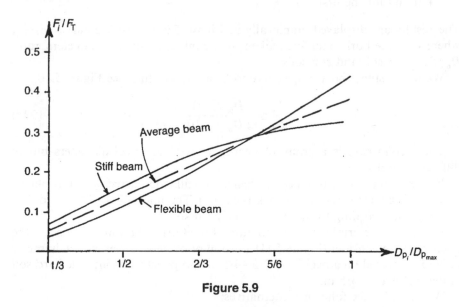

Figure 5.9

The results of the computations show that:

- In the case of average backfill (usually obtained by compaction) on hard soil the forces are distributed nearly equally among the micropiles. As a first approximation we can consider for any micropile a force equal to $F_i = 1 \cdot 2 \, F_T/n$ (n = the number of micropiles).

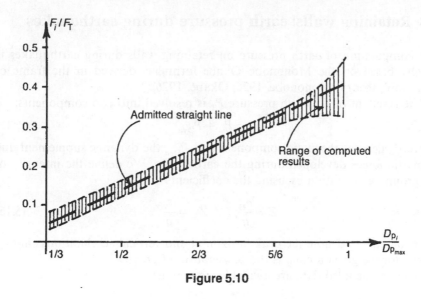

Figure 5.10

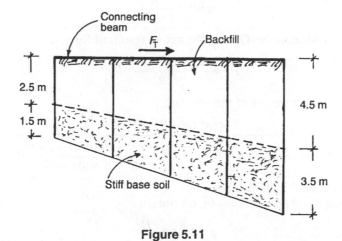

Figure 5.11

- In the case of very soft backfill on a very hard soil the shortest micropiles take $(1 \cdot 6 - 1 \cdot 9) \, F_T/n$, while the longest micropiles take $(0 \cdot 6 - 0 \cdot 9) \, F_T/n$. As a first approximation we can consider for the short micropiles a force equal to $2 \, F_T/n$, whereas for the long micropiles we can consider a force equal to F_T/n.

5.3 Retaining walls: earth pressure during earthquakes

The computation of earth pressure on retaining walls during earthquakes is usually based on the Mononobe–Okabe formulae, derived in the frame of Coulomb's theory (Mononobe, 1926; Okabe, 1926).

The resultant of the earth pressure P_T is resolved into two components:

$$P_T = P_{st} + P_{Dyn} \tag{5.15}$$

where P_{st} denotes the static component and P_{Dyn} the dynamic supplement due to inertia forces developed during the earthquake. We define the intensity of the ground acceleration by using the coefficients

$$Z = \frac{a_H}{g}; \qquad Z_V = \frac{a_V}{g} \tag{5.16}$$

where a_H and a_V denote the peak horizontal and vertical accelerations characteristic for the given area; g is the acceleration of gravity.

When no detailed data are available we may let

$$\frac{Z_V}{Z} \cong \frac{2}{3} \tag{5.17}$$

We define θ by

$$\tan\theta = \frac{Z}{1 - Z_V} \tag{5.18}$$

According to Mononobe–Okabe, the active resultant force is

$$P_T = \frac{K_{a_T}\, \gamma\, H^2 (1 - Z_V)}{2} \tag{5.19}$$

where γ is the unit weight of the soil and

$$K_{a_T} = \frac{\cos^2 (\varphi - \beta - \theta)}{\cos\theta \, \cos^2\beta \, \cos(\delta + \beta + \theta)\left[1 + \sqrt{\left(\dfrac{\sin(\varphi + \delta)\sin(\varphi - i - \theta)}{\cos(\delta + \beta + \theta)\cos(\beta - i)}\right)}\right]^2} \tag{5.20}$$

(for notations see Figure 5.12).

By setting $\theta = 0$ ($Z = 0$, $Z_V = 0$), we obtain

$$P_{st} = \frac{K_{a,st}\, \gamma\, H^2}{2}$$

where

$$K_{a,st} = \frac{\cos^2 (\varphi - \beta)}{\cos^2\beta \, \cos(\delta + \beta)\left[1 + \sqrt{\left(\dfrac{\sin(\varphi + \delta)\sin(\varphi - i)}{\cos(\delta + \beta)\cos(\beta - i)}\right)}\right]^2} \tag{5.21}$$

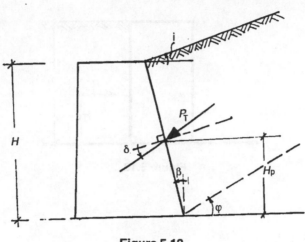

Figure 5.12

The dynamic supplement results in

$$P_{Dyn} = P_T - P_{st} \qquad (5.22)$$

The static component P_{st} acts at a height $H/3$. Experiments performed on gravity (rigid) walls suggest that the dynamic component acts at a height $0 \cdot 6\,H$ (Seed and Whitman, 1970). The resultant will act at the height

$$H_p = \frac{P_{st}\,(H/3) + P_{Dyn}\,(0 \cdot 6\,H)}{P_T} \qquad (5.23)$$

As no similar experiments are available for flexible retaining walls (Dowrick, 1987), we also have to use the equation (5.23) for such type of walls.

A height $H_p \cong 0 \cdot 5\,H$ has been suggested for the resultant force P_T (e.g. AASHTO, 1983).

We note that the increase in the height of the resultant force P_T leads to a significant increase in the overturning moment.

In order to assess the order of magnitude of the increase in the earth pressure due to earthquake, we refer to a simple case where $\beta = i = \delta = 0$ (Figure 5.13) and we let $Z_V/Z \cong \frac{2}{3}$, resulting in

$$P_T = \frac{K_{a.T}\, \gamma\, H^2 (1 - 2\,Z/3)}{2}$$

$$K_{a.T} = \frac{\cos^2(\varphi - \theta)}{\cos^2\theta \left[1 + \sqrt{\left(\dfrac{\sin\varphi\,\sin(\varphi - \theta)}{\cos\theta}\right)}\,\right]^2} \qquad (5.24)$$

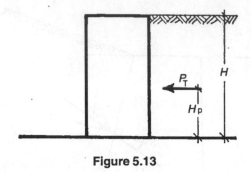

Figure 5.13

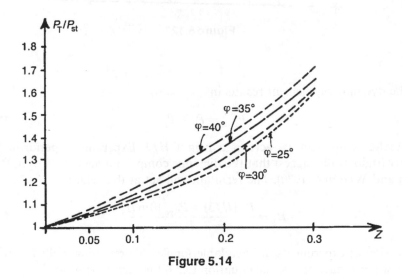

Figure 5.14

and

$$P_{st} = \frac{K_{a.st}\,\gamma\,H^2}{2}$$

$$K_{a.t} = \frac{\cos^2\varphi}{(1 + \sin\varphi)^2} = \tan^2(45^\circ - 0\cdot5\varphi) \qquad (5.25)$$

The graphs shown in Figures 5.14 and 5.15 display the curves P_T/P_{st} and H_P/H versus Z.

These curves can be also applied to approximate P_T/P_{st} for the more general cases (when $\beta \neq 0$, $i \neq 0$, $\delta \neq 0$).

More recent research suggest that a diminished value of the relative acceleration ($Z' < Z$) should be taken into account when using the Mononobe–Okabe

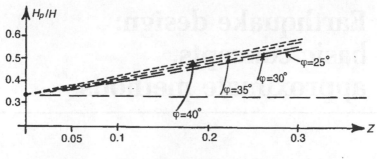

Figure 5.15

formula. The reduced relative acceleration depends on the allowable maximum displacement of the retaining wall; maximum values of $Z' = (0 \cdot 4 - 0 \cdot 5) Z$ have been considered. The main sources for this approach are Richards and Elms (1979), Elms and Richards (1990), the AASHTO (1983) code, the proposed new French code (1990), and Frydman (1992).

Bibliography

Amir, Y. (1989) *Foundation Design in Seismic Areas*, Lishkat Hamehandesim, Tel Aviv (in Hebrew).

Davisson, M. and Salley, J. (1970) Model study of laterally loaded piles. *Journal of the Soil Mechanics and Foundations Division of ASCE*, **96** (SM5), 1605–1628.

Dowrick, D. (1987) *Earthquake Resistant Design*, J. Wiley & Sons, Chichester.

Elms, D. and Richards, R. (1990) Seismic design of retaining walls, in *Design and Performance of Earth Retaining Structures, Proceedings of Conference at Cornell University*, pp. 854–869.

Frydman, S. (1992) Design of retaining walls in seismic areas, in *Proceedings of the Fifth Conference of the Israel Association of Earthquake Engineering*, Tel Aviv, pp. 97–102 (in Hebrew).

Kocsis, P. (1968) *Lateral Loads on Piles*, Bureau of Engineering, Chicago.

Matthewson, M., Wood, J. and Berril, J. (1980) Earth retaining structures. *Bulletin of the New Zealand Society for Earthquake Engineering*, **13** (3), 280–293.

Mononobe, N. (1926) Earthquake-proof construction of masonry dams, in *Proceedings of World Engineering Conference 9*.

Okabe, S. (1926) General theory of earth pressures. *Journal of the Japanese Society of Civil Engineeres*, **12** (1).

Poulos, H. (1979) Group factors for pile-deflection estimation. *Journal of the Geotechnical Division of ASCE*, **105** (GT12), 1489–1509.

Richards, S. and Elms, D. (1979) Seismic behavior of gravity walls. *Journal of the Geotechnical Division of ASCE*, **105** (GT4), 449–464.

Seed, H. and Whitman, R. (1970) Design of earth retaining structures for dynamic loads, in *Proceedings of ASCE Specialty Conference on Lateral Stresses in the Ground and the Design of Earthquake Retaining Structures*, New York, pp. 103–147.

6 Earthquake design: basic concepts, approximate methods

6.1 Evaluation of seismic forces: regular structures

6.1.1 GENERAL DATA, PERFORMANCE CRITERIA, SEISMIC COEFFICIENTS

The performance criteria in seismic design accepted by most of the present codes follow the guidelines established by the Californian Code SEAOC-88. According to these guidelines buildings should:

- resist earthquakes of minor intensity without damage (it is expected that the structure remain in the elastic range);
- resist moderate earthquakes with minor structural and some non-structural damage;
- resist major catastrophic earthquakes without collapse.

For exceptionally important buildings (such as hospitals, fire stations, and power plants) or exceptionally hazardous buildings (such as nuclear power plants), the criteria should be more stringent.

Modern codes evaluate the total horizontal seismic force F_T (equal to the **base shear** V) acting on a building by taking into account the following factors:

1. the **seismic intensity factor,** i.e. the probable intensity of seismic motions, depending on the seismic area in which the building is located (the seismic areas are defined by **seismic maps** included in the codes);
2. the **site factor,** depending on the nature of soil layers overlying the bedrock;
3. the **rigidity factor,** depending on the rigidity of the structural elements and the mass of the building;
4. the **reduction factor,** depending mainly on the ductility of the materials and on the detailing of the structural elements;
5. the **importance factor** of the given building;
6. the total weight W of the building.

In the case of regular structures (i.e. structures that are not very slender and do not display significant irregularities) the seismic forces may be determined according to the **static lateral force procedure,** as detailed in the following.

The total seismic force takes the form

$$F_T = cW \qquad (6.1)$$

where $c = F_T/W$ denotes the **seismic coefficient**. The seismic coefficient is obtained by multiplying several factors:

$$c = \text{(Seismic intensity factor)} \times \text{(Site factor)} \times \text{(Rigidity factor)}$$
$$\times \text{(1/Reduction factor)} \times \text{(Importance factor)}$$

The total seismic force F_T is then distributed to each storey by multiplying it by specified **distribution coefficients** d_i. The force F_i acting at storey i results in

$$F_i = F_T d_i \qquad (6.2)$$

Structures may exhibit 'irregularities', which affect the distribution and magnitude of seismic forces acting on the various resisting elements. These can be classified in two categories:

- irregularities in the horizontal plane of the structure, mainly asymmetry, leading to significant eccentricities and subsequently to torsional forces;
- irregularities in the vertical plane, mainly discontinuities in the rigidities of structural elements and in the masses.

When the structure is very slender or presents significant irregularities, we shall refer to it as a **special structure** and shall determine the seismic forces based on a **dynamic analysis**, usually a **modal analysis** (see section 6.2).

It is important to point out the different approaches in seismic design embodied in US codes (SEAOC-88, UBC-91) and in the European codes. The US codes, unlike most of the European codes, use **factored loads** as well as **strength-reduction factors** in proportioning of structural members; consequently the reduction factors also differ (Luft, 1989).

Therefore, in order to compare the seismic forces obtained from most of the European codes with the seismic forces yielded by the US codes, we have either to multiply the forces resulting from European codes by 1·4–1·5 or, alternatively, to obtain the reduction factors used in most of the European codes by dividing the corresponding US reduction factors by 1·4–1·5.

In the following, we shall refer mainly to the Californian Code SEAOC-88, as a typical modern code.

6.1.2 SEISMIC INTENSITY FACTOR

The size of an earthquake is generally related to the amount of energy released; this is usually measured by the magnitude M defined by Richter (1935) as

$$M = \log A \qquad (6.3)$$

where A denotes the maximum amplitude in μm registered on a standard seismometer placed at a point 100 km from the epicentre (as the existing seismometers are generally located at different distances from the epicentre, equation (6.3) has to be corrected accordingly).

Different procedures are used in seismic design, in order to determine the intensity of the seismic motions, indirectly related to the magnitudes of past earthquakes in the given area. We shall relate to two procedures: the intensity scale, and the predicted horizontal maximum peak acceleration.

(a) Intensity scales

Intensity scales, such as the Modified Mercalli (MM), are based on historical information dealing with the effects of past earthquakes and human feelings. An abridged form of the Modified Mercalli scale is given in the following:

    I. Usually not felt, except under exceptional circumstances.

    II. Felt by persons at rest, especially in upper floors of buildings. Suspended light objects may swing.

    III. Felt indoors. Vibrations sensed like those of a passing car.

    IV. Windows, doors, dishes disturbed. Standing motor cars noticeably rocked.

    V. Felt by nearly everyone, many awakened. Dishes, windows broken. Cracked plaster. Unstable objects overturned.

    VI. Felt by all, many frightened and run outdoors. Some heavy furniture moved. Fallen plaster and damaged chimneys.

    VII. Difficult to stand. Everybody runs outdoors. Moderate damage in well-built structures, major damage in poorly built structures.

    VIII. Slight damage to specially designed structures; substantial damage to ordinary designed structures, with possible partial collapse. Fall of chimneys, factory stacks, columns, monuments and walls. Heavy furniture overturned. Sand and mud ejected, changes in well water.

    IX. General panic. Masonry damaged or destroyed. Considerable damage to specially designed structures, partial or total collapse of ordinary buildings. Buildings shifted off foundations. Conspicuous ground cracking. Underground pipes broken.

    X. Most masonry and framed structures destroyed. Ground severely cracked. Rails slightly bent. Considerable landslides from riverbanks and steep slopes.

    XI. Few, if any, masonry structures remain standing. Bridges destroyed. Broad fissures in ground. Underground pipelines put completely out of service. Earth slumps and land slips in soft ground. Rails greatly bent.

    XII. Damage total. Waves seen on ground surfaces. Objects thrown into the air.

Esteva and Rosenblueth (1964) proposed a relationship between the MM intensity, magnitude M and the epicentral distance; a graphical representation of this relationship is shown in Figure 6.1, following Wakabayashi (1986).

(b) Predicted horizontal maximum peak acceleration

In procedures based on the predicted horizontal ground peak accelerations (a_{max}), usually, the codes refer to a relative acceleration $Z = a_{max}/g$, assumed with a prescribed probability of exceedance (e.g. 10%) within a certain time span (e.g. 50 years); the assessment is based on instrumental and historical regional earthquake data.

The concept of peak ground acceleration (PGA) overlooks several data, important for a correct evaluation of the probable structural damage. The data not taken into account relate especially to the amount of energy released by earthquakes, such as the duration of the acceleration peaks (very short peaks lead to local damage only, without producing extensive collapse) and the number of peak accelerations during a given earthquake (a small number of peaks produce only limited damage). Newmark and Hall (1982) proposed replacing the concept of peak ground acceleration by the more suitable concept of **effective peak ground acceleration**, which now constitutes the basic parameter in the **seismic map** of the Californian code SEAOC-88, and it is probable that it will be also used in the future in other countries.

We should emphasize that consideration of the energetic aspect in evaluating the probable structural damage involves the maximum velocities occurring in earthquakes (see section 6.6); consequently, maps including the probable maximum velocities accompany in the SEAOC-88 code the probable maximum accelerations maps.

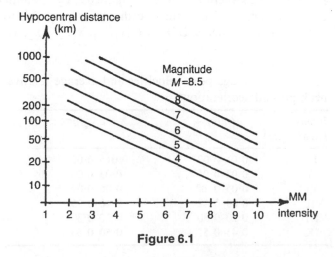

Figure 6.1

Average peak velocities (v_{max}) and average relative peak ground accelerations ($Z = a_{max}/g$) corresponding to the MM scale intensities, as proposed by
Belt (1978), are given in Table 6.1. The values usually prescribed in seismic
codes are $Z = 0 \cdot 05 - 0 \cdot 40$.

6.1.3 SOIL OR SITE FACTOR

The nature and thickness of various soil layers that overlay the bedrock
strongly affect the ground motions at a given site. The phenomenon is quantified by the **site factors** prescribed in seismic codes.

From the site factor point of view, the best soils are the rock-like, as well as
the stiff or dense soils. The most dangerous are soft clays, saturated sediments,
saturated cohesionless soils and loose or recent sediments; sandy soils, where
liquefaction may occur, are particularly dangerous. The site factors usually
prescribed in seismic codes vary between 1 and 1·5, and exceptionally 2.

6.1.4 RIGIDITY AND MASS FACTORS; FUNDAMENTAL PERIOD

The intensity of seismic forces acting on a structure depends on its rigidity. In
order to quantify this effect we have first to define three important dynamic
concepts: degree of freedom, mode of vibration and response spectra.

(a) Degree of freedom

Consider a vibrating structure with lumped masses (Figure 6.2). The degree of
freedom (DOF) is defined as the number of parameters required to determine
the position of the masses at any moment.

Referring to the plane structure shown in Figure 6.2, by assuming a flexible
but inextensible axis, DOF = 3 (e.g. the coordinates x_1, x_2, x_3); by also considering the vertical displacements, DOF = 6 (x_1, y_1, x_2, y_2, x_3, y_3).

Table 6.1 Average peak velocities and average relative
peak ground accelerations

Intensity (MM scale)	v(m/s)	$Z = a_{max}/g$
IV	0·01–0·02	0·015–0·02
V	0·02–0·05	0·03–0·04
VI	0·05–0·08	0·06–0·07
VII	0·08–0·12	0·10–0·15
VIII	0·20–0·30	0·20–0·30
IX	0·45–0·55	0·50–0·55

Referring to the frame shown in Figure 6.3, the masses are generally assumed as lumped at the levels of the slabs (by neglecting the mass of columns and walls); by assuming flexible but inextensible columns and by considering the beams as rigid, DOF = 4.

Note: If a given deformed axis (e.g. a sine curve) is assumed, the structure has a single degree of freedom (SDOF), as a single parameter suffices to determine the position of all the masses at any time. Referring to Figure 6.2, if the coordinate x_1 is known, then x_2, x_3 may be deduced from the condition that the deformed shape is imposd (e.g. a sine curve).

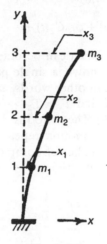

Figure 6.2

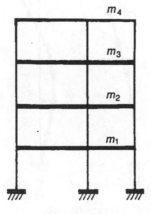

Figure 6.3

(b) Modes of vibration

Consider the vibrating structure shown in Figure 6.4 (DOF = n) due to a short 'perturbation'; let $u_i(t)$ be the deflection of the mass i. The maximum value of $u_i(t)$ is the amplitude ϕ_i. After removing the perturbation the structure vibrates 'freely'; the amplitude decreases with time, owing to damping. It may be shown (see Appendix A1) that the deformed shape of the structure $u_i(t)$ may be obtained as a sum of the amplitudes of n **modes of vibration** $\phi_{i_1}, \phi_{i_2}, \ldots, \phi_{i_n}$, multiplied by the functions $C_1(t), C_2(t), \ldots, C_n(t)$:

$$u_i(t) = u_{i_1}(t) + u_{i_2}(t) + \cdots + u_{i_n}(t)$$

$$= \phi_{i_1} C_1(t) + \phi_{i_2} C_2(t) + \cdots + \phi_{i_n} C_n(t) \tag{6.4}$$

Each mode has a given form, which can be determined by dynamic analysis. As the shape of the mode j is known, a single parameter will suffice to define the positions of all the masses; in other words, each mode is a SDOF system.

Each mode has a **natural period** $T_1 > T_2 > \cdots > T_n$, resulting from the dynamic analysis (see Appendix A). The period is defined as the time that elapses during a 'cycle' (after which the motion repeats itself); the period is usually measured in seconds (s). The number of complete cycles in a unit of time is defined as the frequency of the vibration:

$$f = \frac{1}{T} \tag{6.5}$$

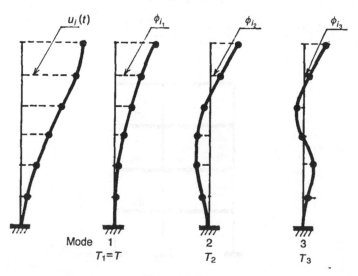

Mode	1	2	3
	$T_1 = T$	T_2	T_3

Figure 6.4

The frequency is measured in Hz $= 1/s$.

The first mode of vibration is the **fundamental mode**, and its period, which is the longest, is the **fundamental period** $T_1 = T$.

(c) Fundamental mode; fundamental period

In the present section, we deal with **regular structures**, i.e. structures that are not very slender and not very irregular (most building structures belong to this category).

For such regular structures, the effect of the first (fundamental) mode is predominant, and the effect of the other (higher) modes can therefore be neglected.

The length of the fundamental period $T_1 = T$ is a criterion adequate to define the type of building from the point of view of its rigidity:

Rigid	$T < 0 \cdot 3\,s$
Moderately rigid	$T = 0 \cdot 3 - 0 \cdot 7\,s$
Slender	$T = 0 \cdot 7 - 1 \cdot 5\,s$
Very slender	$T > 1 \cdot 5\,s$

Buildings with periods larger than 2–3 s are rather uncommon.

An approximate evaluation of the fundamental mode can be obtained by using empirical formulae recommended in the various codes or manuals.

These formulae usually take into consideration the effect of both structural and non-structural elements (facade and partition brickwalls) and therefore result in lower periods (i.e. higher seismic forces) than the 'accurate' methods.

During the first earthquake shocks, the assumption of collaboration between structural and non-structural elements is justified, but towards the end of the earthquake this collaboration is doubtful (if special measures are not taken to ensure it for strong motions); consequently, the seismic forces decrease during the earthquake, tending to those provided by 'accurate' methods (where the structural elements only are taken into account). On the other hand, the **damping ratio** (see Appendix A1) decreases towards the end of the earthquake, and this slows down the rate of decrease of the seismic forces. Using empirical formulae for evaluating the fundamental period is usually the safer course; moreover, these formulae provide more accurate values than the 'accurate methods', at least for the first stage of the earthquake.

This paradox is emphasized in the commentaries to the proposed seismic French code (1990):

The use of empirical formulae is at the same time the simplest and most accurate method of taking into account the stiffness of nonstructural elements.

In the case of moment-resisting frames (Figure 6.5), the fundamental period $T_1 = T$ is given as a function of either the number of storeys (n) or the total height of the structure (H):

$$T = 0 \cdot 1\,n \qquad (6.6)$$

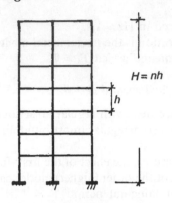

Figure 6.5

or (according to the SEAOC-88 code):

$$T = 0 \cdot 0732 \, H^{3/4} \qquad (6.7)$$

for reinforced concrete moment-resisting frames, or

$$T = 0 \cdot 0854 \, H^{3/4} \qquad (6.8)$$

for steel moment-resisting frames (H in m).
   An average value:

$$T = 0.0793 \, H^{3/4} \qquad (6.9)$$

can be used without significant error.
   Another type of empirical formula takes into account the width L of the building, too e.g. those recommended by the French code (1976) (Figure 6.6):

$$T = \frac{0 \cdot 09 \, H}{\sqrt{L}} \qquad (6.10)$$

for reinforced concrete frames, or

$$T = \frac{0 \cdot 10 \, H}{\sqrt{L}} \qquad (6.11)$$

for steel frames (H and L in m).
   Table 6.2 points out the differences between periods computed according to the above-mentioned empirical formulae (reinforced concrete frames with a width of 20 m and storey heights of 3 m have been considered).
   In the following we shall refer to a procedure for evaluating the fundamental period of multi-storey uniform frames by considering only the structural elements. In a first stage we shall deal with a uniform one-bay frame by assuming that the beams are rigid (Figure 6.7a). The period T can be expressed

Table 6.2 Fundamental periods T(s) for moment-resisting frames, according to various formulae

H	Equation (6.6)	Equation (6.7)	Equation (6.9)	Equation (6.10)
20	0·7	0·7	0·7	0·4
30	1·0	0·9	1·0	0·6
40	1·3	1·2	1·3	0·8
50	1·7	1·4	1·5	1·0
60	2·0	1·6	1·7	1·2

in the form

$$T \cong \varepsilon_n T^\circ$$

where

$$T^\circ = \sqrt{\left(\frac{W H^3}{g E I} \right)} \Bigg\}$$

$$(6.12)$$

and W denotes the total weight of the structure.

The curve of ε_n versus the number of storeys n is shown in Figure 6.8. When the deformability of the beams is taken into account (Figure 6.7b), the period T will increase accordingly. In this case we can use the equation

$$T \cong \varepsilon_n \sqrt{\mu} \, T^\circ \qquad (6.13)$$

The coefficient μ was defined in section 1.2.5 as $\mu = u_{max}/u_R$, where u_{max} denotes the maximum deflection of the considered one-bay frame under a given pattern of horizontal forces and u_R the same deflection when the beams are rigid; a curve of μ versus $v = k_b/k_c$ is shown in Figure 1.21.

We have to point out that the uniform one-bay frame shown in Figure 6.7b may represent the equivalent (substitute) frame of a uniform multi-bay frame (see section 1.3). Consequently, equation (6.13) provides an approximate value of the fundamental period of a uniform multi-storey, multi-bay frame.

Equation (6.13) leads to periods close to those computed by 'accurate methods', as it considers only the effect of the structural elements; it usually yields higher periods than the empirical equations (6.6)–(6.11).

In order to obtain an approximate shape of the fundamental mode we may use the following procedure. We load the given structure with the weights W_i directed horizontally instead of the actual, vertical directions (Figure 6.9): the corresponding elastic line (u_w) is close to the shape of the fundamental mode (the Rayleigh method).

The accuracy of this technique has been checked on a 10-storey building frame (Figure 6.10), where:

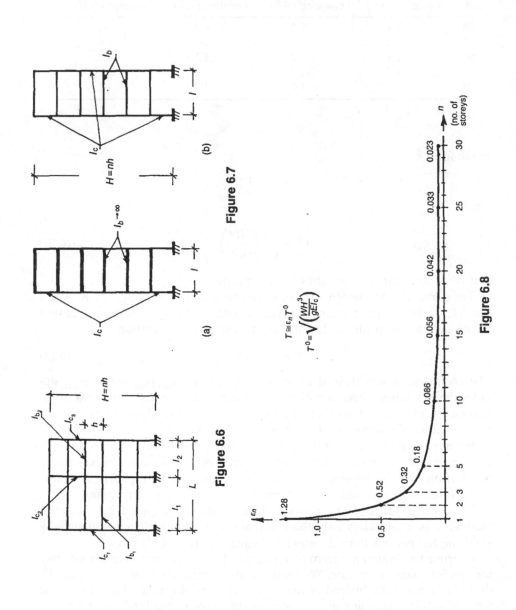

I_b

I_c

$I \bigg|$

$H = nh$

(b)

Figure 6.7

$I_b \to \infty$

I_c

$I \bigg|$

(a)

I_{b_2}

I_{c_3}

h

I_{c_2}

L

I_2

I_{c_1} I_{b_1}

I_1

$H = nh$

Figure 6.6

$T \equiv \varepsilon_n T^0$

$T^0 = \sqrt{\left(\dfrac{WH^3}{gEI_c}\right)}$

ε_n

1.28

1.0

0.5

0.52

0.32

0.18

0.086

0.056

0.042

0.033

0.023

1 2 3   5   10   15   20   25   30

n
(no. of storeys)

Figure 6.8

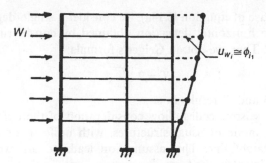

Figure 6.9

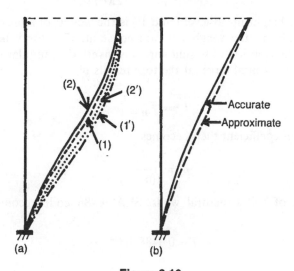

Figure 6.10

(1) = accurate shape for rigid beams;
(1') = approximate shape for the same case;
(2) = accurate shape for very flexible beams;
(2') = approximate shape for the same case.

The greatest differences between deflections occur in the case of rigid beams, but they remain less than 10%, and hence the approximation is justified for practical purposes.

By the same method we may assess the fundamental period:

$$T \cong \frac{2\pi}{\sqrt{g}}\left[\frac{\sum(W_i u_{w_i}^2)}{\sum(W_i u_{w_i})}\right]^{1/2} \tag{6.14a}$$

As a particular case of equation (6.14a), we consider a single-degree-of-freedom system; u_W is the horizontal deflection obtained by considering a horizontal force equal to W. Then we obtain **Geiger's formula**:

$$T = .2 \sqrt{u_W} \tag{6.14b}$$

where u_W is in m and T results in s.

Most modern seismic codes allow consideration of the deformed shape of the fundamental mode of usual structures, with uniform or nearly uniform storeys, as a straight line. This assumption leads to an expression of the distribution coefficient in the form

$$F_i = F_T\, d_i; \qquad d_i = \frac{\sum (W_i\, H_i^2)}{\sum (W_i\, H_i)} \tag{6.15}$$

where H_i is the height up to level i and W_i is the weight of storey i. When the structure is uniform (the weights W_i are constant; the storey heights h_i are constant), the distribution of seismic forces is inverted triangular (Figure 6.11); the maximum horizontal force (at the top) results in

$$F_{max} = \frac{2 F_T}{n + 1} \tag{6.16}$$

The distribution coefficient then becomes

$$d_i = \frac{2 i}{n (n + 1)} \tag{6.17}$$

In the case of RC structural walls, SEAOC-88 code recommends (Figure 6.12)

$$T = 0.0488\, H^{3/4} \tag{6.18}$$

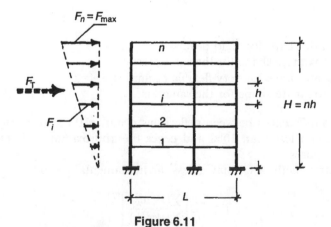

Figure 6.11

whereas the French code (1976) recommends (Figure 6.13)

$$T \cong \left(\frac{0 \cdot 08 \, H}{\sqrt{L}}\right) \sqrt{\left[\frac{H}{L+H}\right]} \tag{6.19}$$

(H, L in m).

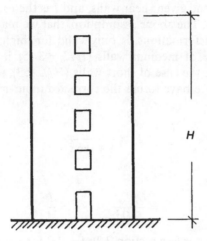

Figure 6.12

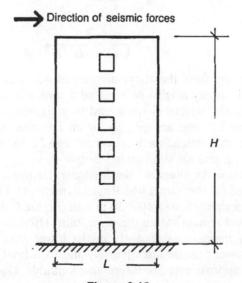

Direction of seismic forces

Figure 6.13

Dynamic analysis for uniform structural walls without openings, by neglecting the effect of shear forces on deformations, yields (see Appendix A1)

$$T = 1 \cdot 787 \sqrt{\left(\frac{W H^3}{g E I}\right)} \qquad (6.20)$$

where $g = 9 \cdot 81$ m s^{-2} is the acceleration of gravity, W is the total weight of the building tributary to the given shear walls, and I is the total moment of inertia of the structural walls. The above assumption, that we may neglect the effect of shear forces on the deformations, is only valid for 'high walls', i.e. for ratios $H/L > 5$. In the case of medium walls ($H/L = 3$–5) it is recommended to consider this effect; in the case of short walls ($H/L < 3$), it is essential.

For this purpose, we have to use the **corrected moment of inertia** I_v instead of I; for instance,

$$I_v \cong \frac{I}{1 + 0 \cdot 6 \, s}$$

where

$$s = \frac{6 f E i^2}{G H^2}; \qquad i^2 = \frac{I}{A} \qquad (6.21)$$

f denotes the **shape factor** (see section 2.2); $f = 1 \cdot 2$ for rectangular sections and 2.2–2.5 for I sections.

In the case of rectangular sections:

$$s = \frac{1 \cdot 41 \, L^2}{H^2}$$

$$I_v \cong \frac{I}{1 + 0 \cdot 85 \, L^2 / H^2} \qquad (6.22)$$

The periods computed from the above-mentioned equations are compared in Table 6.3 (assuming storey heights of 3 m and a constant width of 20 m).

The openings in the structural walls lead to a significant decrease in their rigidity (see section 2.3) and accordingly to an increase in the fundamental period T. For flexible lintels, an increase of 1·6–2·5 may be expected; for average lintels 1·4–2, and for stiff lintels 1·2–1·6.

In order to obtain the shape of the fundamental mode, we may use the procedure described for the frames (the Rayleigh method). The accuracy of this procedure has been checked for a 10×30 m wall (Figure 6.10b) and it has been found that the errors remain within the same limits (10%) in the upper zone of the wall, but may increase significantly in the lower zone ($H/3$). Note that deviations in the lower zone do not adversely affect the final results, so that we may consider the approximate procedure as acceptable. Obviously, for 'squat (short) walls' we have to consider shear deformations.

Table 6.3 Fundamental periods T(s) for structural walls, according to various formulae

H(m)	Equation (6.18)	Equation (6.19)	Equation (6.20)*
20	0·5	0·25	0·4
30	0·6	0·4	0·75
40	0·8	0·6	1·2
50	0·9	0·75	1·55
60	1·0	0·95	2·2

*$W = 42\,000{-}120\,000$ kN and $EI = 6\cdot2 \times 10^7{-}17\cdot4 \times 10^7$ kN m^2 were taken into account.

For dual structures (moment-resisting frames + structural walls), the SEA-OC-88 code recommends equation (6.18). In cases where the frames are more rigid than the structural walls, an average value between equations (6.7) or (6.8) and (6.18) is reasonable.

The effect of soil deformability on the rigidity of structural walls is important; it may decrease their rigidity significantly (see section 2.2.2 for structural walls without openings and section 2.3.7 for coupled structural walls). In section 2.3.7 a procedure is given to compute approximate fundamental periods T in the form $T \cong \sqrt{(T_0^2 + T_r^2)}$, where T_0 denotes the period computed for elastic walls on rigid soil and T_r denotes the same period computed for rigid walls on deformable soil.

6.1.5 DUCTILITY FACTOR; REDUCTION FACTOR

The concept of ductility, developed during the past 40–50 years, has an important role in explaining structural behaviour during earthquakes.

Ductility is defined as a property of the material, element or structure subjected to cyclic loads, to display large inelastic deformations before failure. The most commonly used **ductility ratios** are:

- **displacement ductility**:

$$\mu = \frac{\delta_m}{\delta_y} \tag{6.23}$$

where δ_m is the maximum displacement expected to be attained, and δ_y is the displacement at yield point.

Sometimes δ_m is replaced by the displacement at failure δ_u, although $\delta_m \lessgtr \delta_u$; then

$$\mu \cong \frac{\delta_u}{\delta_y} \tag{6.23a}$$

- **curvature ductility:**

$$\mu = \frac{\varphi_m}{\varphi_y} \cong \frac{\varphi_u}{\varphi_y} \qquad (6.24)$$

where φ_m is the maximum curvature expected to be attained, φ_u is the curvature at failure, and φ_y is the curvature at yield point.

In the following we shall refer to displacement ductility. In the case of RC structures, the reinforcement detailing has a strong effect on ductility.

The range of ductility ratios usually varies between $\mu = 1$ (non-ductile) and $\mu = 6-8$ (high ductility). In special cases, very high ductility ratios of 8–12 can be achieved. 'Ductile structures' have ductility ratios of around 4–5.

The ductility ratios do not appear explicitly as a factor in the evaluation of seismic forces prescribed by the codes, but they are included in the **reduction** (or behaviour) **factors** R. The reduction factors result from comparisons between the results obtained from elasto-plastic or non-linear analysis and elastic, linear analysis.

These factors depend, in addition to the ductility ratio, also on other parameters (e.g. the existence of alternative 'lines of defence', such as lateral force system redundancy and non-structural elements, changed damping and period modification with deformation).

We shall refer first to the dependence of the reduction factor on ductility (by using a **primary reduction factor** R_0), and we shall then give corrected values of the reduction factor R by taking into account other effects. Finally, the reduction factor R thus obtained represents the ability of the structure to sustain strong seismic motions and absorb energy in excess of the allowable stress limit, without collapse.

The primary reduction factor R_0 is defined as

$$R_0 = \frac{F_m}{F_y} \cong \frac{F_u}{F_y} \qquad (6.25)$$

where F_m denotes the force, corresponding to the maximum expected displacement; F_u is the failure load; and F_y is the yield force.

The primary reduction factor R_0 can be determined according to two criteria, the equal maximum displacement and the equal energy, as follows.

(a) The equal maximum displacements criterion (Figure 6.14)

Analyses have been performed by Clough (1970) for 20-storey RC moment-resisting frames, subjected to dynamic forces yielded by given accelerograms, and by Derecho *et al.* (1978) for 20-storey cantilever walls, both types of structures having relatively long natural periods. The computations were performed for two alternatives:

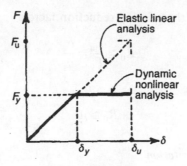

Figure 6.14

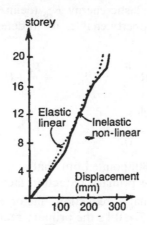

Figure 6.15

- a dynamic inelastic non-linear analysis (leading to maximum horizontal deflections u'_{max});
- an elastic linear analysis (leading to maximum horizontal deflections u''_{max}).

The two analyses yielded nearly equal maximum deflections:

$$u'_{max} \cong u''_{max}$$

These results can be explained by the significant reduction in stiffness evidenced by the non-linear analysis, accompanied by two opposite effects: on the one hand, it leads to an increase in displacements due to lower rigidity; on the other hand, it leads to a decrease in the magnitude of forces due to the longer periods, and thus to smaller displacements. The results of the above-mentioned analyses show that the two opposite effects are nearly equal. This property is illustrated in Figure 6.15.

We now can deduce the primary reduction factor:

$$R_0 \cong \frac{F_u}{F_y} = \frac{\delta_u}{\delta_y}$$

As $\mu = \delta_u/\delta_y$, we deduce

$$R_0 \cong \mu \tag{6.26}$$

(b) The equal energy criterion

Veletsos, Newmark and Chelapati (1965) and Newmark (1970) have shown that, for low and moderate natural periods, we can accept that the elastic energy E_{el} and the elasto-plastic energy E_{ep} (defined in Figure 6.16) are approximately equal. This property enables us to determine the primary reduction factor R_0 as follows:

$$E_{el} = R_0 \delta_y R_0 \frac{F_y}{2}; \qquad E_{ep} = \frac{\delta_y F_y}{2} + (\mu \delta_y - \delta_y) F_y$$

$E_{el} = E_{pl}$ yields

$$R_0 = \sqrt{(2\mu - 1)} \tag{6.27}$$

Chopra and Newmark (1980) pointed out that analysis of several spectra (see section 6.2.2) show that the primary reduction factor $R_0 = \mu$ fits for periods $T > 0.5\,$s, while the primary reduction factor $R_0 = \sqrt{(2\mu - 1)}$ fits for $T = 0.1 – 0.5\,$s; in the range $T < 0.1\,$s the primary reduction factor $R_0 = 1$.

In a few specific structures (e.g. nuclear reactor structures) we have to ensure that the main structural elements remain in the elastic range for all levels of earthquake ground motions. For most structures such a condition is economically prohibitive, and we have to allow for inelastic yielding, on condition that it will not impair the vertical load capacity of the building (see Commentaries to SEAOC-88). This is expressed by assuming conventional

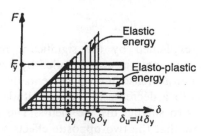

Figure 6.16

seismic forces (F) equal to those expected for elastic, linear design (F_{el}), divided by the reduction factor R:

$$F = \frac{F_{el}}{R} \tag{6.28}$$

The reduction factor R represents, as defined above, a corrected value of the primary reduction factor R_0. Values of the factor R are given in all modern codes, and they usually vary between 3 and 8 (when non-factored seismic loads are considered, such as in most European codes) or between 5 and 12 (when factored seismic loads are considered, such as in the US codes). The values of the reduction factors R were determined on the assumption that the accepted inelastic deformations are distributed quite uniformly in all structural elements.

Note that lessons derived from recent strong earthquakes have shown clearly the much better behaviour of buildings relying on RC structural walls than that of buildings relying on moment-resisting RC frames (Aoyama, 1981; Fintel, 1991, 1994). This can be explained by two main factors (Scarlat 1993):

- Structural walls are usually designed on the assumption of rigid soil (fixed base): this leads to an overestimation of their stiffness and subsequently to their overdesign (see section 3.3).
- Designers neglect the ductility provided by the soil surrounding the foundations of the rocking structural walls; this may significantly change the evaluated ductility of the walls.

Paulay and Priestley (1992) have already noted this aspect, and have stated that

It is now recognized that with proper study, rocking may be an acceptable mode of energy dissipation. In fact, the satisfactory response of some structures in earthquakes can only be attributed to foundation rocking. For rocking mechanisms the wall superstructure and its foundation should be considered as an entity. In this context rocking implies soil–structure interaction.

We have only to add that it is rather surprising that this essential aspect has awakened little interest among researchers and designers. The interesting proposal of Luong (1993) dealing with the concept of energy-dissipating index of soils (EDI) should be noted.

The existing reduction factors (e.g. those prescribed by the SEAOC-88 code) in fact encourage the wide use of ductile moment-resisting frames as preferred earthquake-resistant elements instead of structural walls, and this contradicts lessons of past earthquakes (Fintel, 1994).

We have to determine in design the 'real seismic deflection' (check of drift, pounding, $P\Delta$ effects). Most codes recommend determining the real deflections Δ in the form

$$\Delta = R \Delta_{el} \tag{6.29}$$

where R is the reduction factor prescribed by codes and Δ_{el} is the elastic, linear deflection due to the prescribed seismic forces.

Equation (6.29) overlooks the effect of several reserves, usually neglected in design (e.g. the existence of non-structural elements), thus leading to overestimation of the real deflection by several tens of percent.

The SEAOC-88 code prescribes a computation of the real deflections Δ by dividing Δ_{el} by the factor ($3\ R_w/8$) instead of the reduction factor R_w. By taking into account that in the code factored seismic loads are considered, this prescription provides, in fact, a reduction of displacements computed according to equation (6.29) by several tens of percent, and this is close to the actual conditions encountered.

In concluding, it should be emphasized that the important concept of ductility has been sometimes overstated, and this has led to exaggerated deformability of the structure as a whole, and to dangerous $P\Delta$ effects, which may have been responsible for the collapse of several multi-storey buildings (Eisenberg, 1994); Dowrick (1987) also noted that

until the 1980s, research and codes had rightly been preoccupied with overcoming the excessive brittleness and unreliability of ill-reinforced concrete. However, there may have been too much emphasis on creating ductility for ductility's sake.

6.1.6 IMPORTANCE FACTOR

Seismic codes prescribe importance factors I, depending on the occupancy categories of buildings; the I-factor increases the lateral forces by up to 25–40%. In the SEAOC-88 code the I-factor is limited to 1.25, considering that

the details of design and construction often dominate the seismic performance. Therefore, increasing these aspects for essential facilities can improve performance more effectively than relying solely on increased design force levels...

6.1.7 VERTICAL SEISMIC FORCES

The horizontal seismic forces are accompanied by vertical seismic forces. Their intensity is usually assumed to be $\frac{2}{3}$ of the horizontal forces.

The effect of vertical forces is important for several structural elements:

- Horizontal cantilevers–where the downwards vertical forces must be added. The SEAOC-88 code prescribes checking cantilevers for upwards forces equal to 20% of the vertical loads.
- Prestressed concrete beams–where supplementary vertical seismic forces may disturb the balance between stresses due to vertical loads and prestressing forces. The SEAOC-88 code prescribes checking prestressed beams by considering 'not more than 50% of the dead load, alone or in combination with lateral force effects'.
- Precast elements–where the upwards seismic forces decrease the horizontal friction. On the safe side, it is requested to ignore friction forces when designing connections of precast elements (SEAOC-88).

6.2 Evaluation of seismic forces: special structures

6.2.1 GENERAL APPROACH

In cases where the structure is very slender or has significant irregularities, in either the horizontal or in the vertical plane, we define it as a **special structure** and we have to carry out a **dynamic analysis**. This analysis can be performed by using one of the following procedures:

- modal analysis (see section 6.2.3 and Appendix A);
- direct integration of the equations of motion by a step-by-step technique (a **time history analysis**).

For usual design it suffices to perform a modal analysis: we determine the forces and resultant stresses for each considered mode of vibration according to response spectra given in seismic codes and we superimpose the results 'statistically'. In very special cases we have to perform a time history analysis (usually a non-linear analysis) based on a number of ground motion time histories; the dynamic structural response results from a numerical integration of the equations of motion. The limit of slenderness beyond which a modal analysis is required depends on the magnitude of the fundamental period T (usually a limit of 1·5–2 s is prescribed) or on the total height of the building (usually a limit of 70–80 m is required).

The proposal for the revision of the French code formulated by AFPS (1990) prescribes as a limit the height of 75 m or the period.

$$T = \left(\frac{H}{30}\right)^{3/4} \tag{6.30}$$

which yields $T = 0·74$ s for $H = 20$ m, $T = 1·24$ s for $H = 40$ m, $T = 1·68$ s for $H = 60$ m, and $T = 2$ s for $H = 75$ m.

6.2.2 RESPONSE SPECTRA

During earthquakes, the foundations are subjected to random displacements u_G of the ground (velocity $\dot{u}_G$, acceleration $\ddot{u}_G$). The total displacement u_{TOT} comprises, in addition to the ground displacement u_G, the relative displacement u of the structure (see Appendix A):

$$u_{TOT} = u_G + u; \qquad \dot{u}_{TOT} = \dot{u}_G + \dot{u}; \qquad \ddot{u}_{TOT} = \ddot{u}_G + \ddot{u} \tag{6.31}$$

where an overdot (·) denotes the time derivative.

The basic data for a dynamic analysis are the recorded displacements, velocities or accelerations of the ground during earthquakes (Figure 6.17). From these recordings we may evaluate the response of the simple pendulum shown in Figure 6.18 during a given earthquake.

It is of practical interest to draw diagrams showing the response of pendulums of variable height (i.e. of various rigidities, expressed by their periods T) to the registered ground motions. By considering a number of records in a

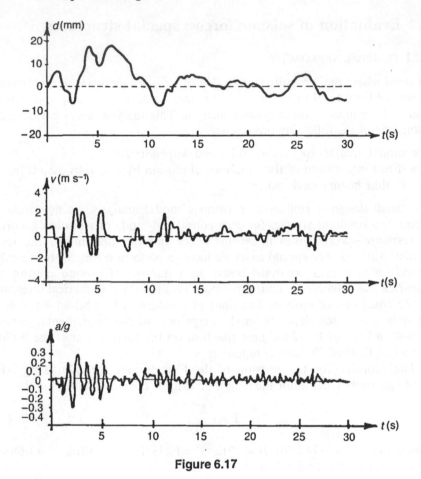

Figure 6.17

given area and retaining the maximum values, we may draw curves exhibiting the variation of the maximum relative displacements (S_d), maximum relative velocities (S_v) and maximum total accelerations (S_a) as functions of the period of the pendulum, as shown in Figure 6.19.

These are the **spectral responses** of the structure to the given earthquake. For each damping ratio ζ there is a different spectrum. Owing to the presence of damping, the peaks of the recordings are smoothed out and the response spectra are significantly flattened. Most codes assume a damping ratio of $\zeta = 0.05$.

The spectra S_d, S_v and S_a are interrelated as follows:

$$S_v = \omega\, S_d; \qquad S_a = \omega\, S_v = \omega^2\, S_d \tag{6.32}$$

where ω denotes the **circular frequency**:

$$\omega = 2\pi f = \frac{2\pi}{T} \tag{6.33}$$

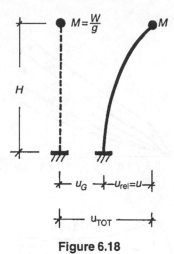

Figure 6.18

The equations (6.32) suggest the possibility of representing a given spectrum and of reading its ordinates ('spectral values') on three different scales, which are usually logarithmic. Two different spectra are drawn in Figure 6.20, one for $\xi = 0$ and the second for $\xi = 0\cdot10$; these are based on a representation of the 1940 E1 Centro earthquake, in direction N–S as drawn by Newmark (1970).

The straight lines represent the maximum ground displacements $(d_{G_{max}})$, velocities $(v_{G_{max}})$ and accelerations $(a_{G_{max}})$.

In the case considered $d_{G_{max}} = 31\,\text{cm}$, $v_{G_{max}} = 32\,\text{cm s}^{-1}$ and $a_{G_{max}} = 0.33\,g$ (g = the acceleration of gravity). Let us consider a point A on the spectrum $\xi = 0\cdot10$, corresponding to the period $T = 0\cdot5\,\text{s}$. We read:

- on the scale S_d, $d_{max} = 3\cdot5\,\text{cm} = 0\cdot11\,d_{G_{max}}$;
- on the scale S_v, $v_{max} = 45\,\text{cm s}^{-1} = 1\cdot40\,v_{G_{max}}$;
- on the scale S_a, $a_{max} = 0\cdot64\,g = 1\cdot94\,a_{G_{max}}$.

That is, an SDOF (pendulum-like) system, with a natural period of $0\cdot5\,\text{s}$, located in El Centro during the given earthquake, would display a maximum relative displacement of $3\cdot5\,\text{cm}$, a maximum relative velocity of $45\,\text{cm s}^{-1}$ and a maximum total acceleration of $0\cdot64\,g$.

Inspection of the response spectra shows an important property: the absolute maximum accelerations of very rigid structures of the considered SDOF system ($T < 0\cdot1\,\text{s}$) approach the ground accelerations (the structures behave like rigid bodies attached to the ground) and the corresponding maximum forces are very high (they tend to $F_{max} = ma_{G_{max}}$, where m denotes the considered mass). The displacements are relatively small. In contrast, the maximum relative displacements of very slender structures ($T > 3\,\text{s}$) are high, and the maximum forces are relatively low (much lower than $m\,a_{G_{max}}$).

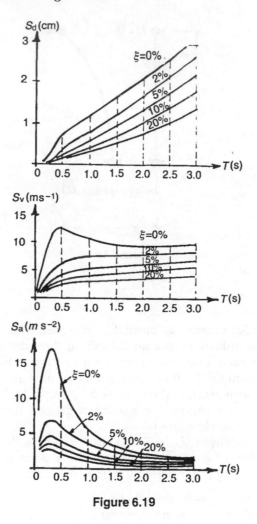

Figure 6.19

We should point out that the type of soil strongly influences the response spectra, and therefore different response spectra have to be drawn.

We distinguish between two types of acceleration spectra. In Figure 6.21a, the peak value of S_a lies in the low periods range: in such cases amplification may occur for rigid structures, and consequently they are dangerous. In Figure 6.21b, the peak value of S_a lies in the high periods range; in such cases, resonance may occur for slender structures and these, too, are consequently dangerous.

In order to allow for both types of earthquake, most modern codes also provide spectra with an enlarged 'plateau' covering rigid and moderately slender structures as well (Figure 6.21c).

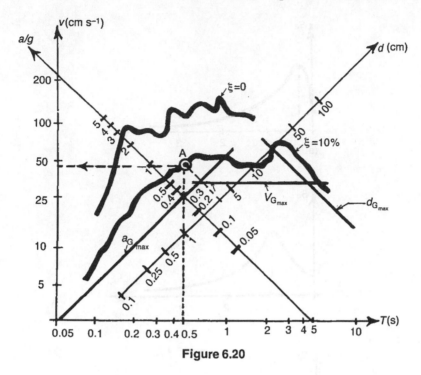

Figure 6.20

6.2.3 MODAL ANALYSIS

It may be shown (see Appendix A) that the vibration modes can be 'decoupled' so that each one can be studied independently. The corresponding modal analysis yields the maximum forces:

$$F_{i_1} = \frac{C_1 \, W_i \, \phi_{i_1}}{T_1^\alpha} \quad \text{(mode 1)}$$

$$F_{i_2} = \frac{C_2 \, W_i \, \phi_{i_2}}{T_2^\alpha} \quad \text{(mode 2)}$$

$$\vdots \qquad\qquad \vdots$$

$$F_{i_j} = \frac{C_j \, W_i \, \phi_{ij}}{T_j^\alpha} \quad \text{(mode } j)$$

$$\vdots \qquad\qquad \vdots$$

$$F_{i_n} = \frac{C_n \, W_i \, \phi_{i_n}}{T_n^\alpha} \quad \text{(mode } n)$$

(6.34)

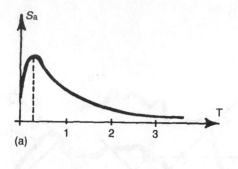

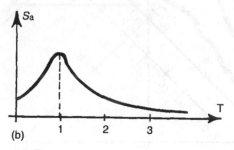

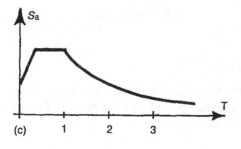

Figure 6.21

where C_j is a constant depending on the seismic zone, the response spectra, including the effect of the type of soil, and the importance of the building; and W_i is the weight acting on the floor i. C_j and the exponent α vary with the codes; SEAOC-88 prescribes $\alpha = \frac{2}{3}$.

$$F_{i_j} = V_j d_{i_j}; \qquad d_{i_j} = \frac{W_i \phi_{i_j}}{\sum(W_i \phi_{i_j})} \qquad (6.35)$$

It follows from equation (6.35) that the forces F_{i_j} (acting at level i, corresponding to mode j) are proportional to the product $W_i \phi_{i_j}$. When the weights are uniform, the forces F_{i_j} follow the shape of the mode j (Figure 6.22).

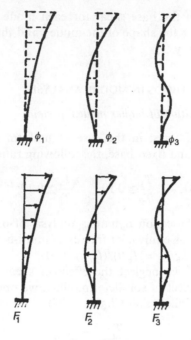

Figure 6.22

The number of modes to be considered in the analysis is defined by various codes, or is chosen by the designer according to the desired degree of accuracy (see Appendix A). Usually, no more than three modes need to be considered; analysis of very slender structures, three-dimensional analysis and special dynamic analysis (involving, for instance, interaction between soil and structure) need more than three modes.

The forces corresponding to mode j lead to the **structural response** r_j (bending moments, shear or axial forces, deflections). Taking into account that the maximum responses are not simultaneous, a **statistical superposition** is usually admitted (see Appendix A1):

$$r_{\text{TOT}} = \sqrt{(r_1^2 + r_2^2 + r_3^2 + \cdots)} \tag{6.36}$$

Two main features of modal analysis are noted as compared with the static lateral force procedure:

- It takes into account the effect of higher modes of vibration, in addition to the fundamental mode; this affects the magnitude of the total lateral force (the base shear).

- The distribution of the base shear for each mode of vibration j is performed according to the shape of the mode j and the values of the masses acting on each storey.

6.2.4 APPROXIMATE METHODS IN MODAL ANALYSIS

(a) Approximate evaluation of higher modes' periods

- Moment-resisting frames: in the case of uniform structures with more than eight storeys and fixed base, the following ratios hold:

$$\frac{T_2}{T_1} \cong 0{\cdot}31; \qquad \frac{T_3}{T_1} \cong 0{\cdot}19; \qquad \frac{T_4}{T_1} \cong 0{\cdot}12; \qquad \frac{T_5}{T_1} \cong 0{\cdot}10$$

These results are based on dynamic analysis of one-bay frames (which may be considered as equivalent frames of multi-bay structures) up to 20 storeys with ratios $k_b/k_c = (I_b/l)/(I_c/h) \geqslant 0{\cdot}1$.

- Structural walls: if we neglect the effect of shear forces (allowable for $H/L > 5$) and the effect of soil deformability, we obtain for uniform walls without openings (Pilkey and Chang, 1978):

$$\frac{T_j}{T_1} = \frac{1{\cdot}425}{(2j-1)^2} \tag{6.37}$$

resulting in:

$$\frac{T_2}{T_1} \cong 0{\cdot}16; \qquad \frac{T_3}{T_1} \cong 0{\cdot}06; \qquad \frac{T_4}{T_1} \cong 0{\cdot}03; \qquad \frac{T_5}{T_1} \cong 0{\cdot}02$$

When we consider the effect of shear forces on the deformations these ratios may change significantly, and they depend on the coefficient $s = 6fEi^2/GH^2$; for $s = 0{\cdot}35$ we obtain

$$\frac{T_2}{T_1} \cong 0{\cdot}24; \qquad \frac{T_3}{T_1} \cong 0{\cdot}11; \qquad \frac{T_4}{T_1} \cong 0{\cdot}08; \qquad \frac{T_5}{T_1} \cong 0{\cdot}06$$

Note that the effect of soil deformability on the rigidity of structural walls is usually important, and overlooking this effect may lead to significant errors (see section 2.2).

In order to determine the order of magnitude of the effect of soil deformability, we consider an RC structural wall with a height of 30 m and a horizontal rectangular section $0{\cdot}20 \times 6{\cdot}00$ m.

Assuming a fixed base, the following ratios are obtained:

$$\frac{T_2}{T_1} \cong 0{\cdot}18; \qquad \frac{T_3}{T_1} \cong 0{\cdot}13; \qquad \frac{T_4}{T_1} \cong 0{\cdot}07; \qquad \frac{T_5}{T_1} \cong 0{\cdot}05$$

When soil deformability is considered:

- for subgrade modulus $k_s = 100\,000\,\text{kN}\,\text{m}^{-3}$:

$$\frac{T_2}{T_1} \cong 0\cdot15; \qquad \frac{T_3}{T_1} \cong 0\cdot09; \qquad \frac{T_4}{T_1} \cong 0\cdot06; \qquad \frac{T_5}{T_1} \cong 0\cdot03$$

- for subgrade modulus $k_s = 20\,000\,\text{kN}\,\text{m}^{-3}$:

$$\frac{T_2}{T_1} \cong 0\cdot10; \qquad \frac{T_3}{T_1} \cong 0\cdot06; \qquad \frac{T_4}{T_1} \cong 0\cdot04; \qquad \frac{T_5}{T_1} \cong 0\cdot02$$

The effect of shear forces was taken into account in all computations.

(b) Approximate assessment of higher modes' shapes

As the effect of the higher modes is usually much smaller than the effect of the first mode, we may use rough approximations regarding their shapes without adversely affecting the final results.

The Spanish code (1976) proposes the schematic shapes displayed in Figure 6.23. Their use, together with approximate evaluations of the periods, enables the effect of higher modes to be taken into account, with no need to resort to modal analysis computations.

(c) Envelopes of 'global effects'

Proposals have been made to construct **envelope diagrams** for bending moments and shear forces, including the effects of higher modes, as functions of the **basic diagrams** (including the effect of the fundamental mode only), multiplied by given factors.

CEB-85 uses the envelopes displayed in Figure 6.24. Numerical examples that we have performed show that the resultant stresses yielded by the CEB-85 envelopes are much higher than those obtained by a regular modal analysis.

The Romanian code (1978) recommends using the envelope shown in Figure 6.25. Numerical examples we have performed show that the coordinate $0\cdot55\,M$ is underestimated.

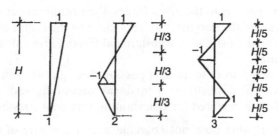

Figure 6.23

The SEAOC-88 code proposes a simpler method to account for the effect of higher modes: in order to increase the shear forces in the top floors (where the effect of higher modes is more significant) a fraction (F_{top}) of the total horizontal force F_T is assumed to be concentrated at the top of the building (Figure 6.26). Then the remaining resultant force $(F_T - F_{top})$ is distributed to each storey. The computations are performed as for a 'regular structure', without referring explicitly to the higher modes.

Capra and Souloumiac (1991) propose to take into account the effect of higher modes by multiplying the resultant stresses due to the fundamental mode by a factor:

$$\rho = 1 + 0 \cdot 05 \left(\frac{T_c}{T_1}\right)^{4/3} > 1 \cdot 05 \tag{6.38}$$

where T_c is defined in Figure 6.27; T_1 denotes the fundamental period.

6.3 Structural irregularities

Structural irregularities are considered as one of the main reasons for the poor seismic performance of buildings. Most codes require special treatment of structures with significant irregularities.

We may classify the irregularities into two main groups: plan irregularities (in the horizontal plane) and vertical irregularities (in the vertical plane).

6.3.1 PLAN IRREGULARITIES

Significant irregularities in the horizontal plane of a structure qualify it as a special structure, and a dynamic analysis (in most cases a modal analysis) is required. We shall distinguish between several kinds of horizontal irregularity: asymmetric structures (the most important one), presence of re-entrant corners, diaphragm discontinuity, and non-parallel structural elements.

(a) In the case of **asymmetric** structures subjected to horizontal forces, the slabs (assumed to be rigid in their planes) describe a displacement that can be resolved into two components, as follows (see section 4.1):

- a translation parallel to the given forces; their resultant passes through the **centre of mass** of the slab (usually very close to the centre of gravity of the slab), associated mostly with **translational forces (stresses)**; a planar analysis can be used;
- a rotation with respect to an axis passing through the **centre of rigidity** of the slab, associated mostly with **torsional forces (stresses)**; a three-dimensional analysis is required for determining torsional stresses.

As the centre of rigidity does not coincide with the centre of mass, a **storey torsion moment** will develop: $M^T = Ve$, where V is the storey shear (the

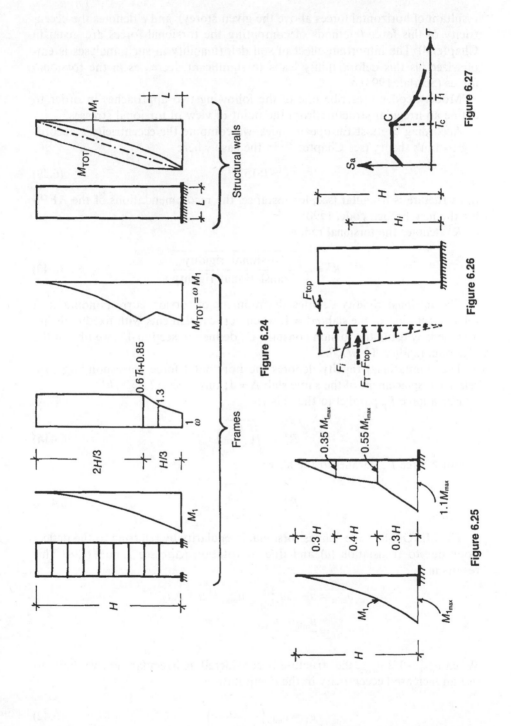

Figure 6.24

Figure 6.25

Figure 6.26

Figure 6.27

resultant of horizontal forces above the given storey), and e denotes the eccentricity of this force (methods of computing the torsional forces are given in Chapter 4). The important effect of soil deformability in such analyses is emphasized, as this deformability leads to significant decreases in the torsional forces (Scarlat, 1993).

Modern codes prescribe one of the following two approaches in order to define an irregular structure from the point of view of torsional stresses.

According to most European codes, we compute the eccentricity e according to Lin's theory (see Chapter 4); in the case where

$$e > 0.15\,R^T \tag{6.39}$$

the structure is irregular (see, for instance, the recommendations of the AFPS for the new French code, 1990).

R^T denotes the **torsional radius**:

$$R^T = \sqrt{\frac{\text{Torsional rigidity}}{\text{Translational rigidity}}} \tag{6.40}$$

The **torsional rigidity** denotes the moment of torsion corresponding to a relative rotation of the slab $\varphi^T = 1$, when vertical elements with fixed ends are assumed; by using the torsion constant C_0 defined in section 4.1, we obtain the torsional rigidity: $12\,E C_0/h^3$.

The **translational rigidity** denotes the horizontal force corresponding to a relative displacement of the same slab $\Delta = 1$; this gives $12\,E\sum I/h^3$.

For a force F_y parallel to the axis y:

$$R_y^T = \sqrt{\left(\frac{C_0}{\sum I_x}\right)} \tag{6.41a}$$

For a force F_x parallel to the axis x:

$$R_x^T = \sqrt{\left(\frac{C_0}{\sum I_y}\right)} \tag{6.41b}$$

The SEAOC-88 code defines torsional irregularity by referring to the deflections due to translation (u) and due to rotation (Δu); see Figure 6.28. This results in

$$u_{min} = u - \Delta u_1; \qquad u_{max} = u + \Delta u_2$$

$$u_{aver} = \frac{u_{min} + u_{max}}{2}$$

When $u_{max} > 1.2\,u_{aver}$, the structure is considered as irregular, and we have to use an increased eccentricity in the computations:

$$e_T = e_{min}\left(\frac{u_{max}}{1.2\,u_{aver}}\right)^2 \tag{6.42}$$

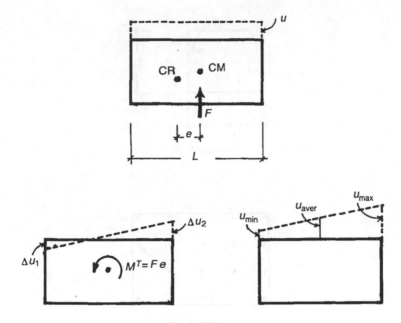

Figure 6.28

$e_{\min}$ is the minimum (accidental) eccentricity prescribed in order to account for uncertainties in the location of the loads. It is usually limited to 5% of the building dimension at the considered level, perpendicular to the direction of the given forces.

In section 4.4 it was shown that asymmetric structures can be classified from the point of view of their plane irregularity by using **torsional index TI**, defined in equation (4.15), into three categories: regular, moderately irregular and significantly irregular.

The torsional index is also used in the 'first screening' of existing buildings (see Chapter 7).

(b) Re-entrant corners (Figure 6.29) lead to local stress concentrations, which may be important when the sizes of the niches are significant; and structure is usually considered irregular when $c_x > 0{\cdot}15\,L_x$ and $c_y > 0{\cdot}15\,L_y$.

(c) When important non-parallel and asymmetric vertical elements are present (Figure 6.30), the structure is considered irregular.

6.3.2 VERTICAL IRREGULARITIES

(a) Weak storey concept

The 'weak storey' concept is related to a discontinuity in the strength of buildings with more than two storeys or higher than 10 m. By denoting by

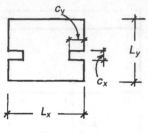

Figure 6.29

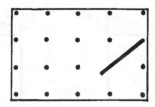

Figure 6.30

$\sum_{(i)} V_u$ the sum of the ultimate shear forces in the vertical elements of the storey i and by $\sum_{(i+1)} V_u$ the same sum for the storey above it, a weak storey is defined as

$$\sum_{(i)} V_u < 0\cdot 8 \sum_{(i+1)} V_u \qquad (6.43)$$

In the case of columns and structural walls without openings we may use the following approximate relation (Figure. 6.31):

$$\sum_{(i)} (bl^2) < 0\cdot 8 \sum_{(i+1)} (bl^2) \qquad (6.44)$$

For profiled structural walls and cores ($\perp, I, \sqsubset, ...$) (Figure 6.32), an approximate relation based on the first stadium of reinforced concrete may be used:

$$\sum_{(i)} \left(\frac{I}{y_{max}}\right) < 0\cdot 8 \sum_{(i+1)} \left(\frac{I}{y_{max}}\right) \qquad (6.45)$$

It has been assumed that the reinforcement ratio of the walls is nearly uniform.

A condition similar to equation (6.43) must be checked for the lintels (also on the assumption that their reinforcement ratio is uniform).

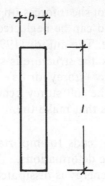

Figure 6.31

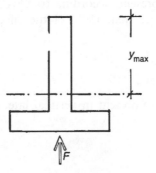

y_{max}

Figure 6.32

(b) Soft storey concept

The 'soft storey' concept is related to a discontinuity in the stiffness of the building.

By denoting by $\sum_{(i)} K_\Delta$ the sum of the translational stiffnesses of the vertical elements of the the storey i and by $\sum_{(i+1)} K_\Delta$ the same sum for the storey above it, the soft storey will occur when

$$\sum_{(i)} K_\Delta < 0.7 \sum_{(i+1)} K_\Delta \qquad (6.46)$$

We may consider an approximate relation:

$$\sum_{(i)} (I_c + I_{sw}) < 0.7 \sum_{(i+1)} (I_c + I_{sw}) \qquad (6.47)$$

I_c (I_{sw}) represent the moments of inertia of columns (structural walls).

It is assumed that the effect of shear forces on the rigidity is nearly the same for both storeys i and $i+1$ and can be neglected in equation (6.47), although the assumption is questionable. The author (1994b) has proposed an alternative way to identify soft storeys: the structure is subjected to a horizontal force at top and we determine the "storey drifts" $\delta_i = u_i - u_{i-1}$ (storey i) and $\delta_{i+1} = u_{i+1} - u_i$ (storey $i+1$); the soft storey occurs when $\delta_{i+1}/\delta_i < 0.7$.

Soft storeys have drawbacks that make them particularly dangerous:

- The vertical discontinuity leads to important stress concentrations, accompanied by large plastic deformations.
- Most of the deformation energy is dissipated by the soft storey columns, and this leads to major overstressing of these elements; onset of plastic hinges may transform the soft storey into a mechanism resulting in collapse.

(c) Mass discontinuity

A mass discontinuity is present, according to SEAOC-88, when 'the effective mass of any storey is more than 150% of the effective mass of an adjacent storey (roofs excepted)'.

(d) In-plane discontinuities

In-plane discontinuities are present in vertical lateral force resisting elements (Figure 6.33).

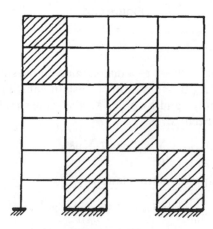

Figure 6.33

(e) Set-back

A discontinuity, as shown in Figure 6.34, leads to a concentration of stresses in the adjacent storeys and to the need for special reinforcement of these zones. The SEAOC-88 code defines the set-back by $B_b > 1 \cdot 3 \ B_t$. The ATC 1978 Model Code proposed a criterion based on rigidities, namely:

$$K_{\Delta_b} > 5 \ K_{\Delta_t} \qquad\qquad (6.48)$$

6.3.3 IRREGULARITIES AND STRUCTURAL DESIGN

Structural irregularities derive from real, often unavoidable, programmatic and aesthetic architectural reasons. It is practically impossible, and usually not necessary, to avoid theses irregularities completely.

However, we have to take into account that lessons from past earthquakes demonstrate beyond all doubt that severe irregularities were among the main causes of building collapse during earthquakes. Therefore we have to take care to identify structural irregularities, and to quantify their damage potential with a reasonable degree of precision. We can then introduce the structural modifications and strengthenings required to ensure adequate seismic resistance.

Significant inelastic deformations develop close to the structural discontinuities during major levels of ground motion, which may lead to local distress and sometimes to general collapse. In this context it should be borne in mind that the values of the reduction factor R (see section 6.1.5) have been determined by assuming that the cyclic inelastic deformations are quite uniform in all structural elements. In the case of irregular structures, these values are not valid any more, and we have to decrease them drastically when designing 'irregular elements', i.e. to increase the forces acting on these elements compared with those based on code formulae.

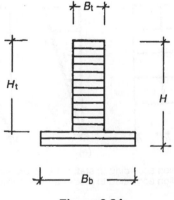

Figure 6.34

In fact, only inelastic dynamic analyses, tests of structures subjected to dynamic cyclic loads, or pertinent observations of structural damage after major earthquakes can provide reliable data for quantifying possible damage caused by various types of structural irregularities. As these are beyond the practical possibilities of usual seismic design we have to be content with prescriptions derived from such analyses, tests or observations and sometimes, unfortunately, also from rules of thumb.

In the following, we shall refer to the prescriptions and recommendations included in the SEAOC-88 code and its commentary.

Two ways are prescribed in this code for dealing with structural irregularities:

- performing a dynamic analysis, usually a modal analysis;
- using **amplification factors** in the design of the 'irregular elements', in order to take into account the stress concentrations specific to structural irregularities.

As shown in section 6.2.3, the modal analysis has two main features as compared with the static lateral forces procedure: it takes into account the effect of higher modes of vibration; and the distribution of the base shear is in agreement with the vibration mode shape. This distribution can provide in several cases a more accurate picture than the linear distribution yielded by the static lateral force procedure. As an example, we refer to the structure shown in Figure 6.35.

Regarding the effect of higher modes, we have to point out that the need to take these modes into account depends only on the rigidity of the structure: these modes do not provide the information needed to quantify the effect of stress concentrations, which accompany structural irregularities.

SEAOC-88 provides several 'penalties' for structural irregularities, as follows.

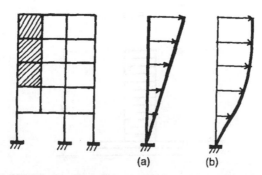

(a) Distribution according to static lateral force procedure
(b) Distribution according to model analysis (mode 1)

Figure 6.35

(a) Plan irregularities

For torsional irregularity, as shown in section 6.31, in the case of siginificant eccentricities we have to increase the torsional forces by increasing the eccentricity resulting from a static analysis (equation (6.42)).

Connections between vertical resisting elements and horizontal diaphragms of structures in seismic zones with relative peak ground accelerations $Z > 0\cdot3$ must be designed by considering allowable stresses without the usual increase of 33%.

(b) Vertical irregularities

Weak storeys must be designed to resist forces obtained by modal analysis, multiplied by a factor of ($3\ R_w/8$). As the average reduction factor R_w prescribed by SEAOC-88 is 8, we have to multiply the design forces by 3. Note that the recommended multiplying factor is independent of the extent of weakness of the considered storey: it is the same for a decrease in strength of 35% or 65%. We point out that amplification factors should be applied also for soft storeys (see section 6.6).

Recommendations for the analysis of set-back buildings are based on the proposals formulated by Blume, Knox and Lindskog (1958).

The buildings are divided into four categories, depending on the ratios H_t/H and B_t/B (Figure 6.34). For each category a specific kind of structural analysis is recommended.

The proposals were formulated before computer programs made dynamic inelastic analyses possible, and their supporting evidence is not available. It would be preferable to propose amplification factors for the design of the structural elements close to the discontinuity level, based on dynamic inelastic analyses.

As an incentive to the design of regular structures, a reduction of the base shear of 10% is permitted for such structures.

6.4 Bounds on the seismic coefficient

6.4.1 GENERAL APPROACH

The seismic coefficient has been defined in section 6.1.1:

$$c = \frac{F_T}{W}$$

where F_T is the total horizontal force, equal to the base shear V, and W is the total weight taken into consideration; W usually includes the dead load and a fraction of the live load.

The seismic coefficient represents the most important parameter in seismic design. It is obtained by multiplying four or five factors; these include the seismic intensity, the soil type, the structural rigidity and its ductility, sometimes taking into account correcting factors due to structural irregularities. None of these factors has a solid scientifically determined basis, and their product is exposed to wide scattering. Therefore, we have to limit the seismic coefficient at both ends: c_{max} in order to avoid overdesign and c_{min} in order to avoid underdesign.

6.4.2 THE UPPER BOUND

There are two methods for determining the upper bound of the seismic coefficient:

- by limiting the values of one or more factors included in the seismic coefficient; for instance, the SEAOC-88 code limits

$$C = \frac{1 \cdot 25 \, S}{T^{2/3}} < 2 \cdot 75 \qquad (6.49)$$

where S denotes the site factor and T the fundamental mode of the structure;
- by limiting the seismic coefficient itself. The Romanian code (1981) limits c to

$$c \leqslant (0 \cdot 30 - 0 \cdot 45) \, Z \qquad (6.50)$$

depending on the type of structure.
Scarlat (1989) proposed:

$$\text{for } Z = 0 \cdot 05: \quad c \leqslant 0 \cdot 10 \, Z$$
$$\text{for } Z = 0 \cdot 30: \quad c \leqslant 0 \cdot 20 \, Z$$

Between these limits we can interpolate:

$$c \leqslant 0 \cdot 08 + 0 \cdot 40 \, Z \qquad (6.54)$$

6.4.3 THE LOWER BOUND

In order to determine a lower limit of the seismic coefficient we can also apply two procedures:

- to limit one or more factors; for instance, SEAOC-88 limits:

$$\frac{C}{R_w} = \frac{1 \cdot 25 \, S}{T^{2/3} \, R_w} \geqslant 0 \cdot 075 \qquad (6.52)$$

where R_w denotes the reduction factor;

- to limit the seismic coefficient itself. Scarlat (1989) proposed the following limits:

$$\text{for } Z = 0 \cdot 05: \quad c \geqslant 0 \cdot 02$$
$$\text{for } Z = 0 \cdot 30: \quad c \geqslant 0 \cdot 05$$

Between these limits we can interpolate:

$$c \geqslant 0 \cdot 014 + 0 \cdot 12\, Z \tag{6.53}$$

It may be useful to translate the seismic coefficients into equivalent wind pressure coefficients. To this end we shall refer to the building shown in Figure 6.36. Assuming an average weight of $3 \cdot 5\, \text{kN m}^{-3}$, the total weight results in

$$W \cong 3 \cdot 5\, BLH$$

The total seismic force:

$$F_s = c \times 3 \cdot 5\, BLH$$

The total wind force:

$$F_w = p_w\, BH$$

where p_w denotes the wind pressure.

Equating $F_s = F_w$ yields

$$p_w = 3 \cdot 5\, c\, L \tag{6.54}$$

(p_w in kN m^{-2}; L in m).

For a width L = 20 m and by admitting

$$p_w = \frac{v^2}{1600}$$

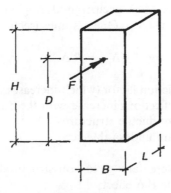

Figure 6.36

where v is the maximum wind velocity (p_w in kN m^{-2}; v in m s^{-1}), we obtain

$$c = 0\cdot01: \quad p_w = 0\cdot70 \text{ kN m}^{-2} \quad v = 33 \text{ m s}^{-1}$$
$$0\cdot02: \qquad\quad 1\cdot40 \qquad\qquad\quad 47$$
$$0\cdot05: \qquad\quad 3\cdot50 \qquad\qquad\quad 75$$

Note that inspection of buildings following the 1906 San Francisco earthquake has shown that buildings designed for wind pressures of $1\cdot5$ kN m^{-2} performed well (Key, 1988). Obviously, several specific factors must be taken into account here, such as the large width of the external walls then used, the important reserves of strength taken by the designer in order to compensate for the poor technical state of the art, and the possibility that the earthquake had a relatively long period – far from the natural fundamental period of the buildings (althouth such periods are not specific for Californian earthquakes). Yet the order of magnitude of the corresponding seismic coefficients remains relevant. We see that a relatively modest seismic coefficient of $0\cdot05$ corresponds to very high wind pressures.

It is interesting to add that in the Bucharest 1977 earthquake (rated at $8-8\cdot5$ on the MM scale), eight-storey buildings designed for seismic coefficients of $0\cdot03-0\cdot04$ suffered little damage, and only three buildings out of several thousand designed for such seismic coefficients collapsed.

The seismic design must not involve excessive horizontal forces, but has to concentrate on other aspects: adequate structural solutions, avoiding excessive irregularities, choosing structures with a reasonably high ductility, good detailing and good workmanship.

6.5 Problems of deformability

6.5.1 GENERAL APPROACH

The plastic deformations (Δ_{pl}) occurring during an earthquake are much greater than the elastic one (Δ_{el}). One assumes (section 6.1.5, equation (6.29)):

$$(\Delta_{pl}) = R \, \Delta_{el}$$

where R denotes the reduction factor (which increases with ductility). It follows that the actual, plastic deflections increase with the ductility; larger deflections are expected in the case of ductile structures.

Excessive deflections are undesirable, as:

- they may lead to severe damage to non-structural elements;
- they may increase the $P\,\Delta$ effect;
- they may cause pounding of adjacent buildings.

6.5.2 STOREY DRIFT LIMITATIONS

SEAOC-88 limits the storey drift to:

$$H < 20 \text{ m}: \quad \frac{\Delta}{h} < \frac{0 \cdot 04}{R_w} \qquad (6.55)$$

$$H > 20 \text{ m}: \quad \frac{\Delta}{h} < \frac{0 \cdot 03}{R_w}$$

We have to bear in mind that the deflections Δ develop as a result of factored loads considered, and that the reduction factors R_w used in the SEAOC-88 are greater than the reduction factors used in Europe.

The CEB 1985 model code distinguishes three situations:

- brittle partitions: $\Delta/h < 0 \cdot 01/R$;
- deformable partitions: $\Delta/h < 0 \cdot 015/R$;   (6.56)
- no connections between partitions and structure: $\Delta/h < 0 \cdot 025/R$.

By considering an average reduction factor $R = 5 \cdot 5$ we obtain

$$\frac{\Delta}{h} = \frac{1}{250} \text{ to } \frac{1}{350}$$

The ratio 1/350 seems exaggerated and difficult to satisfy, especially for moment-resisting frames.

Note: Δ includes translational and torsional effects. Inelastic, non-linear behaviour is taken into account by letting $\Delta = \Delta_{el} R$, where Δ_{el} denotes the storey drift obtained by elastic, linear analysis.

6.5.3 $P \Delta$ EFFECT

Equations of equilibrium are usually formulated for the undeformed structure (first-order analysis). When deflections are excessive and large axial forces exist, we are obliged to take into consideration the supplementary moments and shear forces evidenced by formulating of the equilibrium equations for the deformed structure (second-order theory or $P \Delta$ effect).

Referring to the cantilever shown in Figure 6.37:

First-order theory: $M = F h$
Second-order theory: $M = F h + P\Delta$

As the deflections occurring during earthquakes may be very large, the $P\Delta$ effect can sometimes be important. It can explain the collapse of several multi-storey buildings (Eisenberg, 1994).

According to the SEAOC-88 code we are obliged to consider the $P\Delta$ effect when:

- the storey drift ratio:

$$\frac{\Delta}{h} > \frac{0.02}{R_w} \tag{6.57}$$

- the **stability coefficient** (Figure 6.38):

$$\theta = \frac{P}{V}\frac{\Delta}{h} > 0.10 \tag{6.58}$$

In equations (6.57) and (6.58), Δ denotes the actual, inelastic storey drift. SEAOC-88 recommends computing Δ by multiplying the elastic drift by a factor equal to $3\,R_w/8$. When the $P\Delta$ effect must be accounted for, we have to

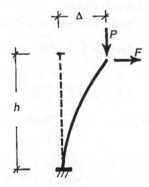

Figure 6.37

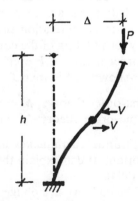

Figure 6.38

design the columns by multiplying the shear force V from the first-order analysis by the stability coefficient θ, defined by equation (6.58).

In high seismicity zones ($Z = 3, 4$), the storey drift limitations usually render the check of the $P\Delta$ effect useless.

6.5.4 POUNDING OF ADJACENT BUILDINGS

Adjacent blocks are usually separated by narrow gaps, which are intended:

- to avoid excessive stresses due to temperature changes and shrinkage of concrete (**temperature joints**);
- to avoid effects of differential foundation settlement (**settlement joints**);
- to avoid excessive torsional forces due to symmetric blocks, especially in seismic areas (**seismic joints**).

As the natural periods of adjacent blocks may differ, seismic motions with different phases may occur, leading to pounding (hammering).

Damage to buildings and local distress can occur, and general collapse due to pounding has been sometimes reported.

We have to distinguish between three different situations, as follows.

1. The slabs of adjacent blocks are located at different elevations; this is the most dangerous situation and may lead to collapse.
2. The slabs of adjacent blocks are located at the same elevations; limited damage can be expected in this case.
3. The same as 2, but strong structural walls/cores are positioned close to the joint (Figure 6.39), perpendicular to the gap line. Only minor damage can be expected in this case.

The standard solution for avoiding the pounding effect is to design sufficiently wide gaps in order to enable each one of the neighbouring blocks to vibrate without contact. It entails widths equal to $d = \Delta_1 + \Delta_2$ at each floor. Δ_1 and Δ_2 are the maximum seismic deflections of the adjacent blocks 1 and 2,

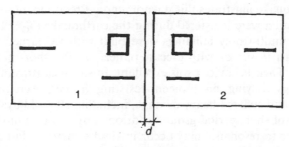

Figure 6.39

where non-linear behaviour must be considered, i.e. a total width of the gap equal to

$$d = (\Delta_{1_{el}} + \Delta_{2_{el}}) R \qquad (6.59)$$

where $\Delta_{1_{el}}$ and $\Delta_{2_{el}}$ are the maximum deflections yielded by an elastic analysis, and R is the reduction factor. The SEAOC-88 code prescribes an amplification factor equal to $(3 R_w/8)$.

By comparing various prescriptions, we reached the conclusion that a global value of the gap width can be admitted in the form

$$d \cong 0.03 \, Z \, H \qquad (6.60)$$

when Δ_1 and Δ_2 are not computed.

We have to recognize that large gaps solve the pounding problem but lead to difficult architectural problems not yet solved satisfactorily; also, we have to ensure that the gaps remain clear of any debris.

An alternative approach, which implies the use of gap elements consisting of a spring and dashpot allowing vibration of adjoining blocks together as a coupled system, is now in the stage of research (Anagnostopoulos, 1988).

Extensive research is being carried out, dealing with behaviour of adjoining structures during pounding (Leibovich, Yankelevsky and Rutenberg, 1994).

6.5.5 EFFECT OF RESONANCE

The accelograms recorded during earthquakes display the predominant periods of the ground motion (T_G). When this period is close to the natural fundamental period of the structure (T), an amplification of displacements and stresses develops (see Appendix A1); this phenomenon is more dangerous when the damping ratio is low $(\xi = 0.01 - 0.02)$.

When we have sufficient earthquake recordings on a given site displaying long period ground motions, we have to design rigid structures, i.e. buildings relying on strong structural walls and cores; the foundations must be as rigid as possible in order to avoid exaggerated rocking effects due to soil deformability. Analysis of seismic behaviour of multi-storey buildings during the 1977 Bucarest earthquake illustrates this phenomenon. Predominant long periods of the ground motion were registered during the earthquake $(T_G = 1 - 1.5\,\text{s})$. Most of the modern multi-storey buildings were built with RC structural walls and cores, either cast *in situ* or with precast panels, having short natural periods $(T = 0.2 - 0.3\,\text{s})$. Their behaviour was excellent. In sharp contrast, the old multi-storey buildings, relying on moment-resisting frames, were more flexible $(T = 0.8 - 1.5\,\text{s})$ and suffered heavy damage, including 30 buildings that collapsed. In the case of short period ground motion, amplification of displacements and stresses due to resonance may occur in rigid structures, but in most cases such structures have enough resistance to avoid collapse. We have to check whether the foundations can avoid excessive rigid rotation of the structures (Figure 6.40), especially in the case of soils exposed to liquefaction.

6.6 Energy approach: application to soft storeys

Most of the modern energy approaches derive from a paper published by Housner in 1956. Further investigations are due to Akiyama (1985) and Bertero (1989).

Housner proposed evaluating the maximum energy absorbed by a structure in the form

$$E = \frac{M \, S_{v,\xi}^2}{2} \tag{6.61}$$

where M is the total mass of the building and $S_{v,\xi}$ is the velocity spectrum for a given damping ratio ξ.

Housner assumed that this energy is constant for a given type of earthquake and a given damping ratio ξ; he based this assumption on the form of the elastic velocity spectra obtained for several earthquakes, in the usual range of rigidities (fundamental periods). In the case of inelastic velocity spectra, usual ductilities, damping ratios and rigidities, the assumption of constant absorbed energy remains practically valid (Veletsos, Newmark and Chelapati, 1965; Aribert and Brozzetti, 1985).

We shall divide the total energy E, absorbed by the structural elements, into elastic energy (E_e) and plastic energy (E_p):

$$E = E_e + E_p; \qquad E_p = E - E_e \tag{6.62}$$

Housner points out that

the energy input is the same when parts of the structure are stressed beyond the elastic limit as it would be if the structure behaved elastically ...

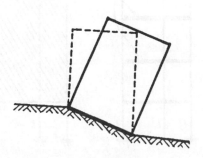

Figure 6.40

This assumption holds true if

the inelastic deformations do not have a major effect on the stiffness characteristics of the structure ...

We emphasize that the results of elasto-plastic time history analyses performed by Clough (1970) on multi-storey frames and by Derecho *et al.* (1978) for RC structural walls have confirmed this basic assumption (section 6.1.5). Finally, we conclude that we can refer to the energy $(E - E_e)$ as a measure of the plastic energy dissipated by the structure.

The author proposed using an energy approach based on Housner's assumption, in order to evaluate the amplification factor to be used in the design of soft storeys (Scarlat, 1994). Two 'extreme models' are considered (Figure 6.41): the first is a perfectly uniform structure with rigid beams, and the second is a structure with rigid storeys, except the soft storey. By comparing the elastic energies stored by both models we deduced the maximum amplification factor c required to ensure that equal plastic energies are dissipated. This factor results in

$$c_0 = \frac{\sqrt{(\sum V_i^2)}}{V} \tag{6.63}$$

where V_i is the shear at storey i and V is the base shear. The amplification factor c_0 represents an upper bound, deduced by considering rigid structures above the ground floor. A reasonable evaluation of the amplification factor c

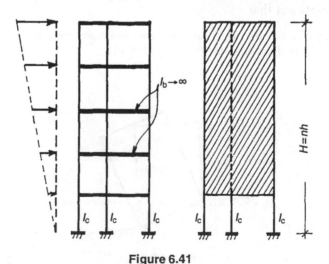

Figure 6.41

can be obtained by interpolation between $c = 1$ (for uniform structures) and $c = c_0$, according to the formula

$$c = c_0 - \frac{(c_0 - 1)K_{GF}}{K_{ST}}$$ (6.64)

where K_{GF} is the lateral stiffness at ground floor and K_{ST} is the lateral stiffness of the structure above ground floor. In the evaluation of the rigidities of RC structural walls, in the case when the aspect ratio $l/H > 1/5$ (H is the total height of the wall), we have to take into consideration the effect of shear forces on the deformations (see section 2.2.1); this can be performed by dividing the moment of inertia of the wall by $(1 + 2s)$ where $s = 6fEI/(GH^2A)$; f is the 'shape factor' of the cross-section of the wall.

6.7 Non-structural elements and non-building structures

6.7.1 NON-STRUCTURAL ELEMENTS

Non-structural elements are defined in this context as architectural components (non-bearing walls, ceilings, parapets, ornamentations) and mechanical-electrical components (such as elevators, sanitary installations and boilers).

Extensive damage to non-structural elements, accompanied by human losses, have been repeatedly reported after major earthquakes. In some cases, the losses in equipment were more important than the building itself.

Dynamic analysis of the behaviour of nonstructural elements is based on **floor response spectra**. The seismic motion of the slab supporting these elements is obtained by a standard dynamic analysis from a given ground motion record; then, from the vibrations determined at the considered slab, the response spectra of the slab can be obtained and the dynamic behaviour of the non-structural elements attached to it analysed accordingly. Such dynamic analyses are directly used only for very special equipment; usually, they are a research tool for determining equivalent static forces to be considered in the design of the non-structural elements and their connections, prescribed in seismic codes.

The SEAOC-88 code requires that non-structural elements be checked as subjected to a seismic force:

$$F_p = ZIC_pW_p$$ (6.65)

where W_p denotes the weight of the equipment and C_p is a coefficient given for various types of non-structural element. For parapets, ornamentations and appendages, chimneys, stacks, signs and billboards, $C_p = 2$. For non-bearing walls and partitions, connections for precast elements and suspended ceilings, as well as for mechanical and electrical equipment, tanks and vessels and bookstacks, $C_p = 0.75$.

The importance factor I varies between 1 and 1·5.

6.7.2 NON-BUILDING STRUCTURES

Non-building structures are designed for horizontal seismic forces computed in the same way as for usual buildings. The formulae used for the fundamental period T as a function of the total height H (section 6.1.4) are not applicable. It is recommended to determine the period T by Rayleigh's method (section 6.1.4). The reduction factor R_w recommended by the SEAOC-88 code varies between 3 and 5. When unfactored seismic loads are used (as in the European codes), reduction factors R between 2 and 3·5 may be considered.

Bibliography

Akiyama, H. (1980) *Earthquake-Resistant Limit-State design for Buildings*, University of Tokyo Press.

Anagnostopoulos, S. (1988) Pounding of buildings in series during earthquakes. *Earthquake Engineering and Structural Dynamics*, **16**, 443–456.

Anon (1982) *The Earthquake of 4th of March 1977 in Romania*, Ed. Academiei (in Romanian).

Aoyama, H. (1981) A method for the evaluation of the seismic capacity of existing RC buildings in Japan. *Bulletin of the NZ National Society for Earthquake Engineering*, **13**(3), 105–130.

Aribert, J. and Brozzetti, J. (1985) Comportement et concepts de dimensionnement des constructions métalliques en zone sismique, in *Génie Parasismique* (ed. V. Davidovici), Presses ENPC, Paris, ch. VII. 5.

Arnold, Ch. (1989) Architectural considerations, in *The Seismic Design Handbook* (ed. F. Naeim), Van Nostrand Reinhold, New York, Ch. 5.

Bertero, V. (1989) Lessons learned from recent catastrophic earthquakes and associated research, in *Proceedings Primera Conferencia Internacional Torroja*, Madrid.

Blume, J. (1958) Structural dynamics in earthquake resistant design. *Journal of the Structural Division of ASCE*, **84** (ST 4), 1–45.

Blume, J., Knox, M. and Lindskog, C. (1958) Proposed setback provisions, Setback Committee.

Blume, J. Newmark, N. and Corning, L. (1961) *Design of Multistory RC Buildings*, Portland Cement Association, Skokie, IL. San Francisco.

Bolt, B. (1978) *Earthquake: A Primer*, W. H. Freeman and Co. San Francisco.

Capra, A. and Davidovici, V. (1980) *Calcul dynamique des structures en zone sismique*, Eyrolles, Paris.

Capra, A. and Souloumiac, R. (1991) Simplified seismic analysis for regular buildings, in *Recent advances in Earthquake Engineering and Structural Dynamics* (ed. V. Davidovici), Presses Académiques, Paris, Ch. IV-7.

Chopra, A. and Newmark, N. (1980) Analysis, in *Design of Earthquake Resistant Structures* (ed. E. Rosenblueth), Pentech Press, London, Ch. 2.

Clough, R. (1970) Earthquake response of structures, in *Earthquake Engineering* (ed. R. Weigel), Prentice-Hall, Englewood Cliffs, pp. 307–334.

Clough, R. and Penzien, J. (1992) *Dynamics of Structures*, McGraw-Hill, New York.

Davidovici, V. (ed.) (1985) *Géne parasismique*, Presses de l'École Nationale des Ponts et Chaussées, Paris.

Derecho, A., Ghosh, S., Iqbal, M., Freskakis, G. and Fintel, M. (1978) Structural walls in earthquake-resistant buildings, dynamic analysis of isolated structural walls–parametric studies. Report to the National Science Foundation, RANN, Constr. Techn. Lab., Portland Cement Association, Skokie, IL, March.

Dowrick, D. (1987) *Earthquake Resistant Design*, J. Wiley, Chichester.

Eisenberg, J. (1994) Lessons of recent large earthquakes and development of concepts and norms for structural design, in *Proceedings of the 17th European Regional Earthquake Engineering Seminar*, Technion Haifa, 1993, Balkema, pp. 29–38.

Esteva, L. and Rosenblueth, E. (1964) Espectros de temblores a distancias moderadas y grandes. *Boletin de la Sociedad Mexicana de Ingineria Sismologica*, 2(1), 1–18.

Fintel, M. (1991) Shear walls–an answer for seismic resistance? *Construction International*, 13(July), 48–53.

Fintel, M. (1994) Lessons from past earthquakes, in *Proceedings of the 17th European Regional Earthquake Engineering Seminar*, Technion, Haifa, 1993, Balkema, Rotterdam, pp. 3–28.

Green, N. (1987) *Earthquake Resistant Building Design and Construction*, Elsevier, New York.

Gupta, A. (1990) *Response Spectrum Method in Seismic Analysis and Design of Structures*, Blackwell Scientific Publications, Boston.

Housner, G. (1956) Limit design of structures to resist earthquakes, in *Proceedings of the First World Conference on Earthquake Engineering*, New Zealand, pp. 5.1–5.13.

Housner, G. (1970) Design spectrum, in *Earthquake Engineering*, Prentice-Hall, Englewood Cliffs, pp. 93–106.

Hudson, D. (1970) Ground motion measurements, in *Earthquake Engineering*, Prentice-Hall, Englewood Cliffs, pp. 107–126.

Key, D. (1988) *Earthquake Design Practice for Buildings*, Thomas Telford, London.

Leibovich, E., Yankelevsky, D. and Rutenberg, A. (1994) Pounding response of adjacent concrete slabs: an experimental study, in *Proceedings of the 17th European Regional Earthquake Engineering Seminar*, Technion, Haifa, 1993, Balkema, Rotterdam, pp. 523–538.

Luft, R. (1989) Comparison among earthquake codes. *Earthquake Spectra*, 5(4), 767–789.

Luong, M. (1993) Energy dissipating properties of soil, in *Proceedings of Second National Earthquake Engineering Conference*, Istanbul, pp. 607–617.

Mazilu, P. (1989) Behaviour of buildings during the 1977 and 1986 earthquakes in Romania, Lecture at Technion, Haifa, November.

Newmark, N. (1970) Current trends in the seismic analysis and design of high-rise structures, in *Earthquake Engineering* (ed. R. Weigel), Prentice-Hall, Englewood Cliffs, pp. 403–424.

Newmark, N. and Hall, W. (1982) *Earthquake Spectra and Design*, Earthquake Engineering Research Institute, Berkeley.

Newmark, N. and Rosenblueth, E. (1971) *Fundamentals of Earthquake Engineering* Prentice-Hall, Englewood Cliffs.

212   Earthquake design

Okamoto, S. (1973) *Introduction to Earthquake Engineering*, University of Tokyo Press.
Paulay, T. and Priestley, M. (1992) *Seismic Design of RC and Masonry Buildings*, J. Wiley & Sons, New York.
Pilkey, W. and Chang Pin Yu (1978) *Modern Formulas for Statics and Dynamics*, McGraw-Hill, New York.
Pironneau, G. (1986) *Seismic Joints: France–USA Workshop*, CNRS, Paris.
Ravin, Sh. (1990) Why buildings are failing during earthquakes? in *Israeli Association of Earthquake Engineering Annual Conference*, pp. 178–184.
Reinhorn, A., Mander, J., Bracci, J. and Kunnath, S. (1989) Seismic damage evaluation of buildings, in *Proceedings of the International Conference on Structural Saftey and Reliability* San Francisco.
Richter, Ch. (1935) An instrumental earthquake magnitude scale. *Bulletin of the Seismological Society of America*, **25**, 1–32.
Rutenberg, A. and Dickman, Y. (1993) Lateral load response of setback shear wall buildings. *Engineering Structures*, **15**(1) 47–54.
Scarlat, A. (1986) *Introduction to Dynamics of Structures*, Institute for Industrial and Building Research, Tel Aviv (in Hebrew).
Scarlat, A. (1989) Critères sismiques dans le projet du bâtiment Shikmona Haifa, in *2ème Coloque National Association Française du Géne Parisismique*, St Rémy-les-Chevreuse, pp. C13/2 7–15.
Scarlat, A. (1993) Soil deformability effect on rigidity-related aspects of multistorey buildings analysis. *Structural Journal of the American Concrete Institute*, **90**(2), 156–162.
Scarlat, A. (1994a) Evaluation of existing buildings for seismic hazard in Israel, in *Proceedings of the 17th European Regional Earthquake Engineering Seminar*, Technion, Haifa, 1993, Balkema, Rotterdam, pp. 481–498.
Scarlat, A. (1994b) Design of soft stories: an energy approach. Lecture given at the Seminarion of the Faculty of Civil Engineering, Technion.
Taranath, B. (1988) *Structural Analysis and Design of Tall Buildings*, McGraw-Hill, New York.
Veletsos, A., Newmark, N. and Chelapati, C. (1965) Deformation spectra for elastic and elastoplastic systems subjected to ground shock and earthquake motions, in *Proceedings of the Third World Conference on Earthquake Engineering*, New Zealand, pp. II 663–674.
Wakabayashi, M. (1986) *Design of Earthquake-Resistant Buildings*, McGraw-Hill, New York.
Wang, T., Bertero, V. and Popov, E. (1975) *Hysteretic Behaviour of RC Framed Walls*, Report EERC, University of California, Berkeley.
Wolf, J. and Skrikerud, P. (1980) Mutual pounding of adjacent structures during earthquakes. *Nuclear Engineering and Design*, **57**, 253–275.
Yarar, R., Eisenberg, J. and Karadogan, F. (1993) A preliminary report on the Erzincan earthquake, March 13 1992, in *Proceedings of Second National Earthquake Engineering Conference*, Istanbul, pp. 452–463.

7 Evaluation of existing buildings for seismic hazard

7.1 Introduction

The sesimic resistance of existing buildings may be evaluated at three levels, according to the required degree of accuracy and to the possibilities offered to the evaluating engineer, as follows.

(a) Level I: Outside inspection (or 'sidewalk survey')

This is designed to be performed from the street, without entry to the building. Some additional information may be provided by insurance companies.

This inspection is aimed at providing only statistical information regarding the seismic vulnerability of a large group of buildings.

(b) Level II: First screening

This is based on visual inspection of the building, accompanied by measurements of certain essential structural members at ground floor level only. We assume that no technical documentation is readily available; some general information can be provided by municipal archives or by direct questioning.

The first screening yields a **seismic index**, which quantifies the approximate seismic resistance of the building and enables a decision to be taken as to the need for further, more accurate investigation.

(c) Level III: Accurate analysis

This is required when:

- the building is of special importance;
- the first screening signals that the structure is probably unsafe for potential seismic hazards.

Such an analysis is usually based on the existing building code, including several specific amendments. A complete, or nearly complete, technical documentation is then required, on the basis of which we can decide whether the building should be strengthened, and determine the main features of the required strengthening work.

7.2 Level I: outside inspection of existing buildings

As noted above, outside inspection can provide only statistical information. It makes possible a very quick – but very approximate – evaluation of buildings. Its cost is obviously low. A description of a typical outside inspection is given in FEMA 154 (1988).

The author has proposed a similar approach, based on the main provisions given in FEMA 154, but adapted to conditions in Israel. In the following we shall give its main features.

The classification of the buildings is based on a **structural score** S:

$$S = S_0 + \Delta S \qquad (7.1)$$

where S_0 is the **basic score** and ΔS are **modifiers**. The basic score S_0 depends on the type of structure and the seismic zone factor Z (defined in the seismic map of the country).

Three seismic zones are considered:

- H, high risk, $Z = 0.25 - 0.30$
- M, moderate risk, $Z = 0.15 - 0.20$
- L, low risk, $Z \leqslant 0.10$

where

$$Z = \frac{\text{Peak ground acceleration}}{\text{Acceleration of gravity}}$$

The proposed values of the basic score S_0 are given in Table 7.1. Where several types of structure are present, the predominant type will be considered. When in doubt, the minimum basic score will be chosen.

The proposed modifiers are identical for all types of structure and all seismical zones (Table 7.2). The structural score $S < 1$ denotes insufficient seismic resistance. $S \geqslant 1$ denotes satisfactory resistance.

☐ **Example 7.1**

A RC frame building, located in a seismic zone $Z = 0.15$ (M), six storeys; poor condition; soft storey; possible pounding (adjacent slabs at same levels); year of construction 1970; soil type S1.

Basic score: $S_0 = 1.5$
Modifiers: $\Delta S = 0$ (medium rise), -0.3 (poor condition), -1 (soft storey), -0.2 (pounding), $+0.5$ (year of construction), 0 (soil).
Structural score: $S = 1.5 + 0 - 0.3 - 1 - 0.2 + 0.5 + 0 = 0.5 < 1$. Insufficient seismic resistance.

Table 7.1   Basic score S_0

Type of structure	Risk category		
	H	M	L
Wood frames	2·2	3	4·2
Steel moment-resisting frames	2	2·5	4·2
Braced steel frames	1·5	2	2·5
Concrete shear walls	2	2·5	3
Precast concrete large panels	1·5	2	2·5
Concrete frames	1	1·5	2
Precast concrete frames	0·5	1	1·5
Reinforced masonry	1·5	1·7	2
Infilled frames	0·7	1	1·5
Plain brick/stone masonry	0·3	0·5	0·7

Table 7.2   Modifiers ΔS

Types of structure	Modifiers ΔS
High-rise buildings (8 storeys or more)	− 0·5
Medium-rise buildings (4–7 storeys)	0
Low-rise buildings (3 storeys or less)	+ 0·3
Poor condition	− 0·3
Poor condition of precast concrete	
structures	− 0·5
Soft storey	− 1
Significant eccentricity	− 0·5
Pounding possible (for medium and high	
rise buildings only)	
Adjacent slabs at same level	− 0·2
Adjacent slabs at different levels	− 0·5
Heavy cladding (precast concrete or	
cut stone)	− 0·5
Short concrete columns	− 0·5
Year of contruction	
Before 1960	− 0·5
1960–1975	0
After 1975	+ 0·5
Type of soil	
S1 (Rock and stiff clay)	0
S2 (Sand, gravel)	− 0·2
S3 (Soft and medium soil or unknown)	− 0·3
S3 + high-rise building	− 0·4

7.3 Level II: first screening of existing buildings

7.3.1 GENERAL DATA

This technique, which originated in Japan (Shiga, 1977; Aoyama, 1981), is based on statistical processing of data dealing with the behaviour of RC structures during strong earthquakes. Subsequently, it was adopted in the USA (Bresler *et al.*, 1977; Hawkins, 1986), China (Chinese Academy of Building Research, 1977) and New Zealand (Glogau, 1980). In 1992, the author proposed a similar technique, extended and adapted to conditions in Israel; it is now being used for checking the seismic vulnerability of buildings in earthquake-prone areas.

The basic assumptions are the following: no technical documentation is available, the structure is checked at ground floor level only, low-quality materials are assumed (for concrete, C17 MPa), and minimum steel ratios are assumed. In a first stage, buildings with four storeys or less were considered; subsequently, data were completed for buildings of up to 12 storeys.

7.3.2 BUILDINGS EXEMPTED FROM FIRST SCREENING

The following buildings located in areas where $Z < 0.2$ are exempted from first screening:

- one-storey buildings with a light roof, where no more than five people are usually present and no materials of special hazard or importance are stored;
- one- or two-storey buildings for offices or dwellings where the ground floor does not exhibit obvious features of a soft storey or large eccentricity;
- buildings up to four storeys where the perimetral structural walls are in reinforced concrete at least 120 mm thick, the length of the RC walls on each side of the perimeter (L_i) covers at least $L_i/2$, and the RC walls reach down to the foundations.

7.3.3 DATA NEEDED FOR COMPUTING THE SEISMIC INDEX

The seismic resistance of a building is quantified by determining a **seismic index** I_s, computed according to the answers to 25 questions included in a questionnaire.

The questions refer to the following aspects:

1. General information (seismic area, year of construction, importance of the building).
2. General description of building, including a sketch (Figure 7.1), number of storeys, data required for determining the total weight of the building above the ground floor.

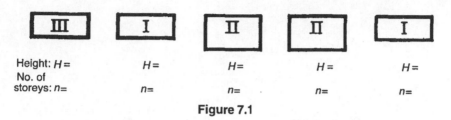

Height: $H =$
No. of
storeys: $n=$ $H=$ $H=$ $H=$ $H=$

$n=$ $n=$ $n=$ $n=$

Figure 7.1

3. Geometric data dealing with the structural elements (Figure 7.2):

(a) For RC shear walls and cores: data required for computing the total horizontal area of the walls A_{sw}; walls at least 120 mm thick and 1 m long are considered. In the case of openings with lintels less than 700 mm high (slab included), the corresponding horizontal areas are neglected. The areas A_{sw} in the main directions L and B are computed separately. The walls of cores are taken into consideration in both directions (L and B). In the case of masonry walls, similar data are required (the horizontal areas A_m for brick masonry and A_{sm} for stone masonry are computed). A minimum thickness of 150 mm for brick masonry and 200 mm for stone masonry is required, as well as a minimum length of 1 m; masonry including windows is considered as without openings, but zones with doors are not taken into consideration (Figure 7.3).

(b) For RC frames, the data for computing the sums $\sum(b_c h_c^2)$ in both directions are required (h_c is the size in the considered direction).

(c) For steel frames the data needed for computing the sums of the elastic section moduli ($W_x = I_x/y_{max}$).

7.3.4 CLASSIFICATION OF STRUCTURES

(a) Plan irregularities

Structures are classified from the point of view of their plan irregularities according to the value of a **torsional index** TI defined in section 4.4 (equation (4.15)). The classification is summarized in Table 7.3.

(b) Vertical regularity

Structures are also classified into three categories from the point of view of their vertical regularity.

In regular structures, each storey is identical or stronger than the storey above it, or the total horizontal area of the RC structural walls of the considered storey (usually the ground floor) $A_{sw} > 1 \cdot 2 Z \sum A/100$ in both main directions L and B; $\sum A$ denotes the total area of the slabs above the considered storey. For details relating to the computation of A_{sw}, see section 2.4.

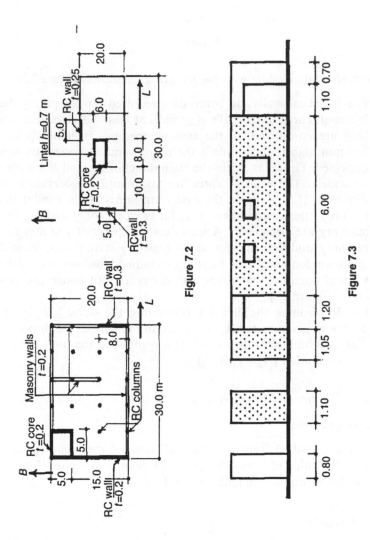

Figure 7.2

Figure 7.3

Table 7.3 Classification of plan irregularities

Regular structures	Symmetric or nearly symmetric in both main directions, or asymmetric with: TI > 1 for structures with RC structural parallel walls in both main directions. TI > 2 for structures with RC structural parallel walls in one direction only.
Moderately irregular structures	$0.5 \leqslant \text{TI} \leqslant 1$ for structures with RC structural parallel walls in both main directions. $1 < \text{TI} < 2$ for structures with RC structural parallel walls in one direction only.
Significantly irregular structures	All other structures.

$\text{TI} = 5\sum(A_{sw} d)/L_{av}\sum A_{sw})$ where A_{sw} is the horizontal area of RC structural wall belonging to a pair of parallel walls; d is the distance between the pair of parallel walls; L_{av} is the average horizontal dimension of the slab above ground floor. For details see section 4.4.

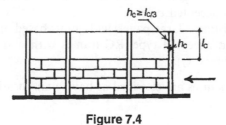

Figure 7.4

A moderately irregular structure occurs when one storey is weaker than the storey above it and

$$\frac{0.6\,Z\sum A}{100} \leqslant A_{sw} \leqslant \frac{1.2\,Z\sum A}{100}$$

All other structures are classified as significantly irregular. When structural short columns (Figure 7.4) supporting vertical loads are present, the structure is classified as significantly irregular.

(c) Pounding of adjacent buildings

Classification of structures from the point of view of pounding of adjacent buildings:

- Regular structures: the adjacent buildings have the same height, seven storeys or less, and the corresponding slabs are located at the same elevation.
- Moderately irregular structures: the corresponding slabs of adjacent buildings are located at the same elevation; adjacent buildings with slabs located at different elevations have three storeys or less.

All other structures are classified as significantly irregular. In cases where the gap between adjacent buildings at each level exceeds $0.03\,Z$ times the corresponding height, the buildings are classified as regular.

(d) Present condition

Classification of structures from the point of view of their present condition: good, satisfactory, unsatisfactory. Criteria are given for the classification; when cracks due to differential settlements are present the condition of the structure is considered as unsatisfactory.

(e) Classification of precast structures

- High resistance: large panels with cast *in situ* horizontal and vertical joints of a known and accepted type.
- Satisfactory resistance: large panels with joints based on welded bars only, of a known and accepted type; RC frames with cast *in situ* joints of a known and accepted type.

All other precast structures are classified as having an unsatisfactory resistance.

(f) Miscellaneous considerations

A special coefficient is reserved for unspecified aspects considered as important by the examiner.

7.3.5 COMPUTATION OF THE SEISMIC INDEX I_s

The seismic index has the form of a safety factor:

$$I_s = \frac{V_a}{V} \tag{7.2}$$

where V is the probable seismic force and V_a is the resisting (allowable) force of the structure.

(a) Seismic force

The seismic force V is determined as

$$V = cW \tag{7.3}$$

where W is the total weight of the building above the ground floor and c is the seismic coefficient, given by

$$c = c° \left(1 + \sum f\right) \tag{7.4}$$

where $c°$ is the **basic seismic coefficient** and f are **modifiers**.

For moment-resisting frames (RC and steel):

$$c° = \frac{1·5\,Z}{\sqrt{H}} \tag{7.5}$$

For other structures:

$$c° = \frac{2·5\,Z}{\sqrt{H}} \tag{7.6}$$

H is the height of the building (in m).

In the case of dual structures, we choose an intermediate value of the basic seismic coefficient, between those yielded by equations (7.5) and (7.6). When the ground floor is a 'soft storey' the corresponding shear force V is multiplied by a factor f_v. The basic seismic coefficients $c°$ were determined as average values corresponding to the provisions of the SEAOC-88 code. The 'modifiers' f depend on the importance of the building, the type of foundation soil and on the degree of horizontal and vertical irregularities; they vary between $-0·2$ and $+0·3$; f_v varies between 1 and 3.

Upper and lower bounds of the seismic coefficients were established as follows:

In seismic areas where $Z = 0·05$: $c = 2-10\%$

In seismic areas where $Z = 0·30$: $c = 5-20\%$

Data for a quick computation of the total weight W are provided.

(b) Resisting (allowable) seismic force

The resisting (allowable) seismic force

$$V_a = V_a° \left(1 + \sum a\right) \tag{7.7}$$

The **basic allowable force** $V_a^?$ is determined as follows (Scarlat, 1994):

- RC structural walls and cores (see section 2.4):

$$V_a = A_{sw}\,\tau_a \tag{7.8}$$

A_{sw} denotes the total horizontal area of structural walls/cores at ground floor level in a given direction: $A_{sw} = \sum L_w\,t$.

Several restrictions imposed on the considered walls are detailed in section 4.4.

τ_a depends on the number of stories (n):

$$\tau_a = \frac{1\cdot44\,p}{\sqrt{n}} \tag{7.9}$$

By assuming a vertical load of $p = 10$ kN m$^{-2} = 0\cdot01$ MPa:

$$\tau_a = \frac{1\cdot44}{\sqrt{n}} \geqslant 0\cdot4 \text{ (MPa)} \tag{7.10}$$

- RC frames (see section 1.4.2):

$$V_{a.c}' = 0\cdot1\frac{\sum(b_c\,h_c^2)f_{ck}}{\varepsilon\,h} \tag{7.11}$$

where f_{ck} is the characteristic (cube) strength of the concrete ($f_{ck} = 17$ MPa is assumed); b_c, h_c are the dimensions of the columns at ground floor level (h_c parallel to the given forces); $\varepsilon = 0\cdot7$ for regular beams and $\varepsilon = 1$ for slab beams; and h is the storey height.

- Steel frames (see section 1.4.3):

$$V_{a.st} = 1\cdot4\frac{\sigma_a^M\sum W_x}{h} \tag{7.12}$$

where $\sum W_x$ denotes the sum of moduli of resistance of the columns at ground floor level, and $\sigma_a^M = 130$ MPa $= 130\,000$ kN m^{-2}.

- Masonry (see section 2.7):

$$V_{a.m}^? = A_m\,\tau_a \tag{7.13}$$

where A_m is the total horizontal area of the masonry at ground floor level in the given direction. A number of restrictions imposed on the considered walls are detailed in section 7.3.3.

The recommended values for τ_a are as follows: plain masonry, $\tau_a = 0\cdot05$ MPa for solid bricks and $0\cdot03$ MPa for hollowed bricks; infilled frames (masonry infill), $\tau_a = 0\cdot2$ MPa for solid bricks and $0\cdot1$ MPa for hollowed bricks.

- Reinforced masonry:

$$\tau_a = \frac{5f_y}{10^4}\frac{L_m}{H} \tag{7.14}$$

(see section 2.7.3)
- Stone masonry:

$$V^\circ_{a,sm} = A_{sm}\tau_a \tag{7.15}$$

where $\tau_a = 0.05$–0.10 MPa.

The 'modifiers' a depend on the year the building was constructed, its present condition, the type of the structure and the foundations type (in cases where no information is available we use the minimum value). These vary between -0.5 and $+0.2$.

(c) Limit-state design

In cases where the structure includes several types of substructures we resort to a limit-state design. The total design force takes the form

$$V_{a,T} = \sum \alpha_i V_{a_i} \tag{7.16}$$

where the **participation factors** α are determined by considering the different ductilities of each substructure. The following factors were chosen, by referring mainly to data provided by Aoyama (1981) and Hawkins (1986) and to the results of our own computations. The proposals set forth by Aoyama (1981) are based on Newmark's equal energy criterion (see section 6.1.5): the **primary reduction factor** R_0 results in:

$$R_0 = \frac{F_u}{F_y} = \sqrt{(2\mu - 1)} \tag{7.19}$$

where the maximum ductility

$$\mu \cong \frac{\delta_u}{\delta_y} \tag{7.20}$$

By admitting a maximum force reduction factor for moment-resisting frames in steel we obtain the participation factors as follows:

- Moment-resisting frames – steel: $\alpha = 1$
  – RC: $\alpha = 0.9$
- Coupled structural walls/cores – RC: $\alpha = 0.85$
- Structural RC walls without openings: $\alpha = 0.8$
- Reinforced masonry: $\alpha = 0.6$
- Infilled frames: $\alpha = 0.4$
- Plain brick and stone masonry: $\alpha = 0.3$

7.3.6 CLASSIFICATION OF BUILDINGS

A building can be classified based on its seismic index I_s into one of the following categories:

1. $I_s > 1\cdot3$: a more accurate check is not required;
2. $1\cdot1 < I_s \leqslant 1\cdot3$: a more accurate check is not urgently required;
3. $0\cdot9 \leqslant I_s \leqslant 1\cdot1$: a more accurate check (according to the seismic code – see section 7.4) is required;
4. $I_s < 0\cdot9$: a more accurate check is urgently required; strengthening of the building is probably needed.

7.4 Level III: accurate analysis of existing buildings

As stated in the introduction, an accurate analysis of an existing building is required either when the building is of special importance, or when the result of the first screening is unsatisfactory.

7.4.1 EXISTING DOCUMENTATION

The ideal situation implies the existence of complete technical documentation, including.

- 'as built' plans of all possible structural elements (including masonry walls) and foundations with all corresponding details;
- in cases where alterations of the original structure were made, the plans and details of these alterations;
- information regarding the quality of construction materials;
- a geotechnical report comprising data on the foundation soil.

Often, such complete documentation is not available and we have to restore it, at least partly, by preparing schematic as-built drawings, including sketches of the main details for the main structural elements.

The minimum data required are as follows:

- Geometric data of all possible structural elements (including masonry walls), their dimensions and positions.
- For RC elements, reinforcement details for the main members: typical columns, structural walls, typical beams and slabs; their identification may be performed either by uncovering the reinforcing bars or by non-destructive techniques (e.g. ultrasound). Concrete quality may be checked by rebound hammer test (a minimum of 10 points per member) and, in the case of unreliable results, by testing drilled samples.
- For steel elements, their dimensions and positions, as well as joint details.

- For brick masonry, whether the masonry is plain or is a part of infilled frames; in the latter case we must check the dimensions of the RC or steel columns and ascertain whether the brick infill is properly connected to the main structural elements. We have to establish the type of the bricks too, and check the quality of mortar (specifically, if the mortar can or cannot be snapped away from the joints by hand with a metal tool).
- For precast structural elements, information regarding the dimensions and reinforcement of the main elements and details of their joints, too.
- Information as to the nature of the foundation soil, either from the original geotechnical report or by executing additional drillings.
- Information as to the foundations (type, depth), by special excavations or by non-destructive techniques.
- Data as to the type of flooring, the type of facade and partitions, in order to evaluate the permanent load.

We define two categories of documentation:

(a) good;
(b) satisfactory.

7.4.2 PRESENT CONDITION OF THE STRUCTURE

The effective capacity of the existing structure is affected by its present condition. We classify the structures from this point of view into three categories:

a. good;
b. satisfactory;
c. unsatisfactory.

The evaluating engineer will choose the proper category by taking into account the following criteria:

- age of the building (benchmark years are determined mainly according to the publication of code revisions);
- alterations to the structural elements;
- damage due to fire;
- cracks due to differential settlements (generally diagonal cracks, visible on both sides of the walls);
- presence of visible deterioration of concrete elements due to corrosion of reinforcing bars and following spalling of the covering concrete; spalling of concrete as a result of aggressive environment; visible segregation of concrete;
- visible rusting or corrosion of steel elements;
- poor quality of mortar in masonry walls.

When the present condition of the structure is considered by the evaluating engineer as very dangerous he will propose either strengthening or demolition of the building or a part of it, without further examination.

7.4.3 PRECAST STRUCTURAL ELEMENTS

Precast elements are classified into four categories:

   a. large panels with wet joints;
   b. large panels with dry joints;
   c. linear elements (columns and beams) with wet joints;
   d. linear elements with dry joints.

The dry joints are based mainly on welded reinforcing bars or steel plates. Wet joints comprise spliced reinforcement bars (by welding or overlapping) included in concrete zones, having a transverse section at least equal to the transverse section of the joining elements and a minimum length of 200 mm.

   The definition of the present condition of precast structures depends mainly on the presence of cracks or spalling of concrete in the joint zones.

7.4.4 EFFECT OF POUNDING (HAMMERING)

When adjacent buildings are separated by expansion joints that are insufficiently wide, pounding can occur, leading to possible damage and sometimes structural failure.

   Pounding is particularly dangerous in situations where the adjacent buildings correspond to the following definitions.

   1. The floors of the adjacent buildings are not at the same elevations.
   2. They have different heights, rigidities or masses.
   3. At least one of the adjacent structures relies only on moment-resisting frames, or the RC walls are not able to absorb the effect of pounding (they are not perpendicular to the expansion joint and close to it).

In order to quantify the effect of pounding we shall refer to the following categories.

   (a) The adjacent buildings correspond to definitions 1, 2 and 3.
   (b) The adjacent buildings correspond to definitions 1 and 2 or 1 and 3.
   (c) The adjacent buildings do not correspond to any of the definitions 1, 2, or 3.

In cases where the gap between adjacent buildings at each floor is more than $0.03\,Z$ times the corresponding height, the building will be considered as belonging to category (c).

7.4.5 STRUCTURAL ANALYSIS

In principle, the structural analysis will be performed according to the methods and specifications included in the existing seismic codes for earthquake design, subject to the recommendations and modifications detailed in the present section.

The structural layout will include, according to the decision of the evaluating engineer, all possible structural elements (including masonry walls) or only a part of them. Plain masonry walls or masonry infill not connected to the main structural elements will not be taken into account for buildings with three stories or more and for buildings in areas with moderate or high seismic activity (the relative ground peak acceleration $Z > 0.15$).

Restrictions detailed in section 7.3.3 for masonry walls are valid for the present section, too.

The force distribution among various resisting elements will be based on elastic analysis in which soil deformability is taken into account; realistic elastic properties of the soil should be used.

In special cases, an inelastic dynamic analysis is recommended, based on simulated earthquakes.

A lower safety factor is usually allowed when checking existing structures, owing to two main factors: a minimum of resistance of the structure (to vertical loads, wind pressure and past earthquakes, even mild ones) has already been demonstrated; and the cost of strengthening an existing building is relatively high when compared with the cost of increasing the seismic strength of a building in the design stage.

The simplest way to consider this reduction consists in multiplying the code-specified forces by given factors. These vary widely according to the sources.

FEMA (1988) recommends values of 0.67 for structures with long periods and 0.85 for structures with short periods; in Romania values of 0.50–0.70 are recommended (Agent, 1994). We recommend a reduction factor of 0.70.

The forces F acting upon the structure are

$$F = 0.7 \, F°(1 + m_d + m_c + m_p + m_h) \tag{7.21}$$

where $F°$ represents the forces given by the code for the design of new structures, and m are 'modifiers' as follows:

- m_d, modifier to include the effect of the type of documentation. According to the classification defined in section 7.4.1: category (a), $m_d = 0$; category (b), $m_d = 0.1$.
- m_c, modifier to include the effect of the present condition of the structural elements. According to the classification defined in section 7.4.3: category (a), $m_c = 0$; category (b), $m_c = 0.1$; category (c), $m_c = 0.2$.
- m_p, modifier for structures with precast elements, only. According to the classification defined in section 7.4.3: category (a), $m_p = 0$; category (b), $m_p = 0.1$; category (c), $m_p = 0.2$; category (d), $m_p = 0.3$.
- m_h, modifier to include the effect of pounding. According to the classification defined at point 4: category (a), $m_h = 0.2$; category (b), $m_h = 0.1$; category (c), $m_h = 0$.

The base shear (total seismic force) $V = \sum F$ is limited by the following values:

$$\text{in seismic zones where } Z = 0.05, \quad V = (0.02 - 0.10)\,W$$
$$\text{in seismic zones where } Z = 0.30, \quad V = (0.05 - 0.20)\,W$$

where W is the total weight of the building.

For intermediate values of Z, we shall interpolate between the aforementioned limits.

The allowable stresses (forces) acting on the existing foundations can be increased with respect to the allowable stresses taken into account in the design of new structures by the following percentages: spread footings subjected to normal forces only, 50%; when subjected to normal forces and moments, 75%; piles, 50%.

7.4.6 NON-STRUCTURAL ELEMENTS

Non-structural elements that by their failure may cause loss of life or injury or damage to equipment considered very important by the owner must be included as a part of the overall building evaluation.

Special attention must be given to masonry walls that carry vertical loads (slabs) and are not properly connected to the main structural members.

The checking of non-structural elements, except masonry walls, will be performed by considering seismic forces:

$$F_p = Z I C_p W \tag{7.22}$$

where W is the weight of the element, and $C_p = 2$.

In the case of masonry walls we shall use the equations shown in section 7.3.5.

Bibliography

Agent, R. (1994) Guiding principles in the strengthening design of earthquake damaged buildings in Romania and case implementations, in *Proceedings of the 17th European Regional Earthquake Engineering Seminar*, Haifa, 1993, Balkema, Rotterdam, 1994, pp. 417–430.

Anon (1982) *The 1977 March 4 earthquake in Romania*, Ed. Acad. RSR.

Aoyama, H. (1981) A method for the evaluation of the seismic capacity of existing RC buildings in Japan. *Bulletin of the New Zealand National Society for Earthquake Engineering*, **13**(3), 105–130.

Bresler, B., Okada, T. and Zisling, D. (1977) Evaluation of earthquake safety and of hazard abatement. UCB/EERC, 77/06, Earthquake Engineering Research Center, College of Engineering, University of California, Berkeley.

Council on Tall Buildings and Urban Habitat (1978) Structural design of tall concrete and masonry buildings, in *Design of Masonry Structures* (ed. A. Hendry, R. Dikkers and A. Yorkdale), American Society of Civil Engineers, Ch. CB-13.

Dowrick, D. (1987) *Earthquake Resistant Design*, 2nd edn. McGraw-Hill, New York.
Englekirk, R. and Hart, G. (1982,1984) *Earthquake Design of Masonry Buildings*, vol. 1,2, Prentice-Hall, Englewood Cliffs.
Ergunay, O. and Gulkan, P. (1990) Earthquake vulnerability, loss and risk assessment: National report of Turkey, *Proceedings Second Workshop on Earthquake Vulnerability*, Trieste, December, pp. 1–45.
Fintel, M. (1991) Shearwalls – an answer for seismic resistance? *Construction International*, 13 (July), 48–53.
Glogau, O. (1980) Low rise RC buildings of limited ductility. *Bulletin of the New Zealand National Society for Earthquake Engineering*, 13(2), 182–193.
Hart, G. (1989) Seismic design of masonry structures, in *The Seismic Design Handbook* (ed. F. Naeim), Van Nostrand, New York, Ch. 10.
Hawkins, M. (1986) *Seismic Design for Existing Structures*, Seminar course manual ACI-SCM-14, Part II, pp. 1–27.
Hendry, A. (1990) *Structural Masonry*, Macmillan, London.
Paulay, T. and Priestley, M. (1992) *Seismic Design of Reinforced Concrete and Masonry Buildings*, J. Wiley & Sons, New York.
Scarlat, A. (1993a) Asymmetric multistory structures subject to seismic loads, Contribution to the evaluation of torsional forces, in *Proceedings Second National Earthquake Engineering Conference*, Istanbul, pp. 30–39.
Scarlat, A. (1993b) Diagonstique préliminaire de la vulnerabilité des bâtiments existants en Israel, AFPS Troisieme Colloque National, St Rémy, pp. TA 57–68.
Scarlat, A. (1994) Evaluation of existing buildings for seismic hazard in Israel, in *Proceedings of the 17th European Regional Earthquake Engineering Seminar*, Technion, Haifa, 1993, Balkema, Rotterdam, pp. 481–498.
Shiga, T. (1977) Earthquake damage and the amount of walls in RC buildings, in *Proceedings Sixth World Conference on Earthquake Engineering*, New Delhi, pp. 2467–2470.
Stafford Smith, B. and Carter, C. (1969) A method of analysis for infilled frames. *Proceedings of the Institution of Civil Engineers*, London, Part 2, **44**, 31–37.
Stafford Smith, B. and Coull, A. (1991) *Tall Buildings Structures: Analysis and Design*, J. Wiley & Sons, New York.
Wakabayashi, M. (1986) *Design of Earthquake-Resistant Buildings*, McGraw-Hill, New York.
Wood, S. (1991) Performance of RC buildings during the 1985 Chile earthquake: implications for the design of structural walls. *Earthquake Spectra*, **7**(4), 607–638.

Postscript

1 Choice of structural solution

The choice of structural solution must take into account the seismicity of the area. It is a nonsense to prescribe the same solutions for Scandinavia and Armenia.

Limiting the general deformability is usually beneficial for multi-storey buildings. It solves *ipso facto* several important problems: the $P \Delta$ effect, pounding, and the integrity of non-structural elements.

Buildings with more than five storeys (in high-seismicity areas) or 10 storeys (in low-seismicity areas) relying on structural walls/cores at all levels have the best chances of survival in major earthquakes.

Symmetric buildings are desirable, but pairs of strong structural walls in both main directions offer a better chance of resisting significant torsional stresses.

High redundancy and the entailing 'invisible reserves of strength' are beneficial. We have to think twice before separating the non-structural masonry walls from the main structural elements in order to avoid asymmetry, so much more that the needed details to achieve this are at best questionable. It is better to ensure good connections with the main structural elements and to design adequate pairs of structural walls in both main directions, able to resist torsional forces.

Local stress concentrations due to vertical irregularities lead usually to limited distress. In contrast, soft storeys are very dangerous in that they have to dissipate most of the energy released by the ground motions; the onset of plastic hinges in the vertical elements of soft storeys may lead to total collapse.

Pounding of adjacent blocks of the building is undesirable, but it is usually accompanied by local distress only. The exception is adjacent structures with slabs at different elevations. The best solution to avoid damage due to pounding is to avoid joints when they are not absolutely necessary to avoid excessive lengths or a very asymmetric configuration of the building.

Lessons from past earthquakes show that in high-seismicity areas, precast elements in the form of large panels with 'wet connections' can perform as well as monolithic structures.

2 Choice of structural design method

When deciding on the degree of accuracy of the structural analysis we have to consider the poor state of the basic supporting data. It is useless, and even ridiculous, to try to replace a lack of basic seismic data (maximum acceler-ations, frequency content of ground motions, filtering effect of soil layers, etc.) by sophisticated, cumbersome and time-consuming dynamic non-linear ana-lyses. They are useful and often necessary in research but not in design (except some very special or very important structures).

Each seismic design must be checked by approximate techniques. To this end, first screening methods are very useful tools.

Seismic codes must be short, simple and clear. These conditions are at least as important as being 'correct'. In this context we should remember that each new major earthquake provides 'structural surprises', so that the term 'correct' must be accepted with a pinch of salt. A long, exhaustive and complicated code suffers the worst possible fate: it will be ignored.

Neglecting soil deformability in dual structures entails completely distorted pictures of stresses when rigidity aspects are involved: it leads to exaggerated rigidities of structural walls and cores and consequently to significant errors in the distribution of seismic forces, in the evaluation of torsional forces and in the assessment of stresses due to temperature changes. Analyses based on this assumption are often an exercise in futility.

At present, we have no reliable methods to assess the 'hidden ductility' of structural walls/cores due to soil–structure interaction during rocking vibra-tions. The current analyses that neglect this effect are at best questionable.

The lack of sufficiently simple procedures for considering soil deformability, as well as other factors governing the real rigidity of structural elements, underscores the need for a new approach in seismic design, based on limit-design criteria. In order to make such techniques available for routine design, some aspects have to be clarified (e.g. limitation of plastic regions in structural elements subject to seismic forces, consideration of torsional forces).

Last but not least we have to remember that earthquake engineering is yet in its infancy. The repeated appeals to 'sound engineering judgement' are a symptom of this fact.

Competent analyses of the effects of major earthquakes, continuous con-frontation of existing theories with these results, as well as progress in geo-physics, are needed to enlarge our basic data, on which we can build more exact methods of seismic design.

The present state of the art in earthquake engineering is best described by a Japanese saying: 'Every error is a treasure. In the discovery of imperfection lies the chance for improvement'.

Appendix A
Structural dynamics: main formulae

A.1 Single-degree-of-freedom (SDOF) systems

A.1.1 UNDAMPED FREE VIBRATIONS

See Figure A.1. Assume initial conditions: $u_0 = 0$; $\dot{u}_0 = 0$.

$$u = -M\ddot{u}\delta$$

Solution: $u = u_{\max} \sin \omega t = (\dot{u}_0/\omega) \sin \omega t$.

$$\omega^2 = \frac{1}{M\delta} = \frac{k}{M} \tag{A.1}$$

(ω = circular frequency in rad s^{-1})

$$f = \frac{\omega}{2\pi} \tag{A.2}$$

(f = natural frequency in Hz = cycles s^{-1}).

$$T = \frac{1}{f} = \frac{2\pi}{\omega} \tag{A.3}$$

(T = natural period in s cycle^{-1} = Hz^{-1}); see Figure A.2.

Approximately (Figure A.3):

$$T \cong 2\sqrt{u_{\mathrm{w}}} \tag{A.4}$$

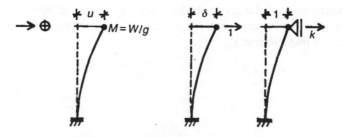

Figure A.1

(u_w in m; T in s): Geiger's formula.

Forces of inertia:

$$F = M\omega^2 u; \qquad F_{max} = M\omega^2 u_{max} \tag{A.5}$$

A.1.2 DAMPED FREE VIBRATIONS

See Figure A.4. Viscous damping:

$$F_{damp} = -b\dot{u} \tag{A.6}$$

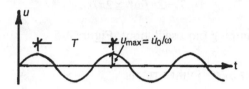

Figure A.2

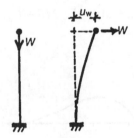

Figure A.3

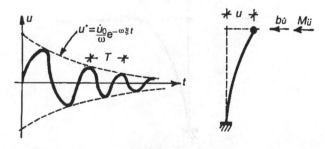

Figure A.4

Damping ratio:

$$\xi = \frac{b}{2\,M\omega} \tag{A.7}$$

Usually: $\xi = 0.02-0.10$.

By assuming $\omega\sqrt{(1-\xi^2)} = \omega^* \cong \omega$:

$$u = \left(\frac{\dot{u}_0}{\omega}\right) e^{-\omega\xi t} \sin \omega t$$

$$T = 2\pi/\omega^* \cong 2\pi/\omega. \tag{A.8}$$

If $\xi > 1$: overdamping (no oscillation)–Figure A.5.

A.1.3 UNDAMPED FORCED VIBRATIONS

See Figure A.6. Exciting force:

$$F(t) = F_0 f(t)$$

$$u = u(t) = [-M\ddot{u} + F_0 f(t)]\delta$$

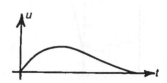

Figure A.5

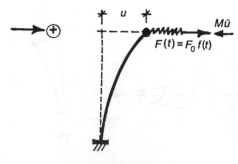

Figure A.6

(a) Harmonic excitation

$$f(t) = \cos pt$$

p = circular frequency of the harmonic excitation in rad s^{-1}.

$$u = u_\omega + u_p$$

$$u_\omega = \left(\frac{\dot{u}_0}{\omega} - \frac{p\,\Delta p}{\omega} \frac{1}{1-(p^2/\omega^2)} \right) \sin \omega t \qquad (A.9)$$

$$u_p = \frac{\Delta p}{1-(p^2/\omega^2)} \cos pt; \qquad \Delta_p = F_0 \delta = \frac{F_0}{M\omega^2}$$

When $p = \omega$ ... resonance $(u \rightarrow \infty)$.

(b) Arbitrary excitation: $p(\tau) = F_0 f(\tau)$

$$u = \frac{F_0}{M\omega} \mathscr{I}(t); \qquad \mathscr{I}(t) = \int_0^t f(\tau) \sin \omega(t-\tau)\,d\tau \qquad (A.10)$$

in which $\mathscr{I}(t)$ is Duhamel's integral.
   Dynamic amplification factor:

$$\mu(t) = \frac{u}{\Delta_p} = \omega \, \mathscr{I}(t) \qquad (A.11)$$

A.1.4 DAMPED FORCED VIBRATIONS

See Figure A.7.

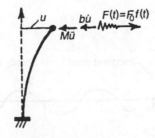

Figure A.7

(a) Harmonic excitation

$$\ddot{u} + 2\omega\dot{u} + \omega^2 u = \Delta_p \cos(pt)$$

$$u = u_\omega + u_p$$

$$u_\omega = e^{-\omega\xi t}(C_1 \cos\omega^* t + C_2 \sin\omega^* t); \qquad u_p = B\cos(pt - \lambda)$$

$$B = \frac{\Delta_p}{\sqrt{[(1 - p^2/\omega^2)^2 + 4p^2\xi^2/\omega^2]}}$$

$$\tan\lambda = \frac{2p\omega\cdot\xi}{\omega^2 - p^2}$$

After a short lapse of time: $e^{-\omega t} \to 0$; $u_\omega \to 0$.

$$u \cong u_p = B\cos(pt - a)$$

('steady state vibrations': Figure A.8).
 Dynamic amplification factor:

$$\mu(t, \xi) = \frac{u}{\Delta_p} \cong \frac{u_p}{\Delta_p} = \frac{\cos(pt - \lambda)}{\sqrt{[(1 - p^2/\omega^2)^2 + 4p^2\xi^2/\omega^2]}}$$

For $\cos(pt - \lambda) = 1$ (maximum):

$$\mu_{\max} = \frac{1}{\sqrt{[(1 - p^2/\omega^2)^2 + 4p^2\xi^2/\omega^2]}} \qquad (A.12)$$

Resonance: $p = \omega$; $\mu = 1/2\xi$ (e.g. for $\xi = 0{\cdot}2$, $\mu = 2{\cdot}5$). $\qquad$ (A.13)

(b) Arbitrary excitation

$$F(\tau) = F_0 f(\tau)$$

Assuming $\omega^* \cong \omega$:

$$u = \frac{F_0 \cdot \mathscr{I}(t, \xi)}{M\omega}; \qquad \mathscr{I}(t, \xi) = \int_0^t e^{-\omega\xi(t - \tau)} f(\tau) \sin\omega(t - \tau)d\tau$$

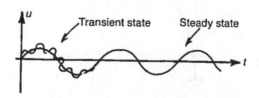

Figure A.8

Dynamic amplification factor:

$$\mu(t,\xi) = \frac{u}{\Delta_p} \cong \frac{u_p}{\Delta_p} = \omega.\mathcal{I}(t,\xi) \qquad (A.14)$$

A.2 Multi-degree-of-freedom (MDOF) systems

A.2.1 UNDAMPED FREE VIBRATIONS: FLEXIBILITY FORMULATION

See Figure A.9.

$$u = u(x,t) = \phi(x)Y(t) \qquad (A.15)$$

Computation of circular frequencies/periods:

$$\begin{vmatrix} (M_1\delta_{11} - \frac{1}{\omega^2}) & M_2\delta_{12} & M_3\delta_{13} & \cdots & M_n\delta_{1n} \\ M_1\delta_{21} & (M_2\delta_{22} - \frac{1}{\omega^2}) & M_3\delta_{23} & \cdots & M_n\delta_{2n} \\ \vdots & & & & \vdots \\ M_1\delta_{n1} & M_2\delta_{n2} & M_3\delta_{n3} & \cdots & (M_n\delta_{nn} - \frac{1}{\omega^2}) \end{vmatrix} = 0 \qquad (A.16)$$

Solution: $\omega_1 < \omega_2 < \cdots < \omega_n$; $T = 2\pi/\omega$: $T_1 > T_2 > \cdots > T_n$.
Computation of coordinates ϕ_{i_j} corresponding to the mode of vibration j:

$$\left. \begin{aligned} (M_1\delta_{11}\omega_j^2 - 1)\phi_{1_j} + M_2\delta_{12}\omega_j^2\phi_{2_j} + M_3\delta_{13}\omega_j^2\phi_{3_j} + \cdots + M_n\delta_{1n}\omega_j^2\phi_{n_j} = 0 \\ M_1\delta_{21}\omega_j^2\phi_{1_j} + (M_2\delta_{22}\omega_j^2 - 1)\phi_{2_j} + M_3\delta_{23}\omega_j^2\phi_{3_j} + \cdots + M_n\delta_{2n}\omega_j^2\phi_{n_j} = 0 \\ \vdots \qquad\qquad\qquad\qquad\qquad\qquad\qquad\qquad \\ M_1\delta_{n1}\omega_j^2\phi_{1_j} + M_2\delta_{n2}\omega_j^2\phi_{2_j} + M_3\delta_{n3}\omega_j^2\phi_{3_j} + \cdots + (M_n\delta_{nn}\omega_j^2 - 1)\phi_{n_j} = 0 \end{aligned} \right\}$$
$$(A.17)$$

We choose an arbitrary value for one of the coordinates, e.g. $\phi_n = 1$, and deduce accordingly ϕ_1, ϕ_2,... from the set of equations; thus we obtain the **modal shapes** $1, 2, ..., j, ..., n$ (Figure A.10).

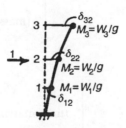

Figure A.9

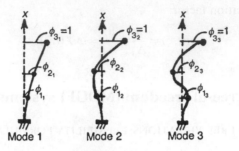

Figure A.10

The final shape of the vibrating elastic line will be obtained by superimposing the modal shapes multiplied by the corresponding functions $Y(t)$:

$$u_i = u_{i_1} + u_{i_2} + \cdots + u_{i_j} + \cdots + u_{i_n}$$
$$= \phi_{i_1} Y_1(t) + \phi_{i_2} Y_2(t) + \cdots + \phi_{i_n} Y_n(t)$$

Forces:

$$F_i = F_{i_1} + F_{i_2} + \cdots + F_{i_j} + \cdots + F_{i_n}$$
$$= M_1 \omega_1^2 \phi_{i_1} Y(t) + M_2 \omega_2^2 \phi_{i_2} Y_2(t) + \cdots + M_n \omega_n^2 \phi_{i_n} Y_n(t)$$

In matrix formulation:

$$F = \begin{bmatrix} \delta_{11} & \delta_{12} & \cdots & \delta_{1n} \\ \delta_{21} & \delta_{22} & \cdots & \delta_{2n} \\ \vdots & & & \\ \delta_{n1} & \delta_{n2} & \cdots & \delta_{nn} \end{bmatrix} \quad \text{(flexibility matrix)}$$

$$M = \begin{bmatrix} M_1 & & & \\ & M_2 & & \\ & & \ddots & \\ & & & M_n \end{bmatrix} \quad \text{(mass matrix)}$$

$$I = \begin{bmatrix} 1 & & & \\ & 1 & & \\ & & \ddots & \\ & & & 1 \end{bmatrix} \quad \text{(unit matrix)}$$

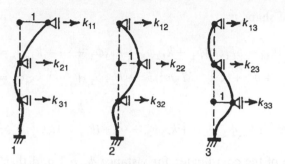

Figure A.11

$$\boldsymbol{\phi}_j = \begin{bmatrix} \phi_{1_j} \\ \phi_{2_j} \\ \vdots \\ \phi_{n_j} \end{bmatrix} \quad \text{(vector of coordinates, mode } j\text{)}$$

Computation of circular frequencies $\omega_1, \omega_2, \ldots, \omega_n$:

$$\det[\boldsymbol{D} - (1/\omega^2) \cdot \boldsymbol{I}] = 0; \qquad \boldsymbol{D} = \boldsymbol{F} \cdot \boldsymbol{M} \tag{A.18}$$

Computation of coordinates ϕ_j corresponding to the mode of vibration j:

$$[\boldsymbol{D} - (1/\omega_j^2)\boldsymbol{I}] \cdot \boldsymbol{\phi}_j = 0 \tag{A.19}$$

A.2.2 UNDAMPED FREE VIBRATIONS: STIFFNESS FORMULATION

See Figure A.11.

Computation of circular frequencies ω:

$$\begin{bmatrix} (k_{11} - M_1\omega^2) & k_{12} & k_{13} & \cdots & k_{1n} \\ k_{21} & (k_{22} - M_2\omega^2) & k_{23} & \cdots & k_{2n} \\ \vdots & & & & \vdots \\ k_{n1} & k_{n2} & k_{n3} & \cdots & (k_{nn} - M_n\omega^2) \end{bmatrix} = 0 \tag{A.20}$$

Computation of shape j:

$$\left.\begin{array}{l} (k_{11} - M_1\omega_j^2)\phi_{1_j} + k_{12}\phi_{2_j} + k_{13}\phi_{3_j} + \cdots + k_{1n}\phi_{n_j} = 0 \\[4pt] k_{21} + \phi_{1_j} + (k_{22} - M_2\omega_j^2)\phi_{2_j} + k_{23}\phi_{3_j} + \cdots + k_{2n}\phi_{n_j} = 0 \\[4pt] \vdots \\[4pt] k_{n1}\phi_{1_j} + k_{n2}\phi_{2_j} + \cdots + k_{n3}\phi_{3_j} + \cdots + (k_{nn} - M_n\omega_j^2)\phi_{n_j} = 0 \end{array}\right\} \qquad (\Lambda.21)$$

We choose one of the coordinates, for instance $\phi_{n_j} = 1$ and then compute ϕ_{1_j}, $\phi_{2_j}, \ldots$.

(c) Matrix formulation

$$K = \begin{bmatrix} k_{11} & k_{12} & \cdots & k_{1n} \\ k_{21} & k_{22} & \cdots & k_{2n} \\ \vdots & & & \vdots \\ k_{n1} & k_{n2} & \cdots & k_{nn} \end{bmatrix} = 0 \quad \text{(stiffness matrix)}; \; K \cdot F = I$$

$$\det(K - \omega^2 \cdot M) = 0 \qquad (A.22)$$

Computation of coordinates ϕ_{i_j}:

$$(K - \omega_j^2 M) \cdot \phi_j = 0 \qquad (A.23)$$

A.2.3 ORTHOGONALITY OF MODES h AND j $(h \neq j)$

Displacements:

$$M_1\phi_{1_h}\phi_{1_j} + M_2\phi_{2_h}\phi_{2_j} + \cdots + M_n\phi_{n_h}\phi_{n_j} = 0 \qquad (A.24)$$

In matrix form:

$$\phi_h^T \cdot M\phi_j = 0 \qquad (A.25)$$

Rigidities:

$$\left.\begin{array}{l} \phi_{1_h}(k_{11}\phi_{1_j} + \cdots + k_{1n}\phi_{n_j}) + \phi_{2_h}(k_{21}\phi_{1_j} + \cdots + k_{2n}\phi_{n_j}) + \cdots \\[4pt] + \phi_{n_h}(k_{n1}\phi_{1_j} + \cdots + k_{nn}\phi_{n_j}) = 0 \end{array}\right\} \qquad (A.26)$$

In matrix form:

$$\phi_h^T \cdot K\phi_j = 0 \qquad (A.27)$$

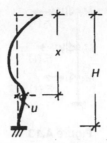

Figure A.12

A.2.4 UNDAMPED FREE VIBRATIONS: UNIFORM CANTILEVERS

See Figure A.12.

Uniform mass: $m = W/(gH)$, where W is the total weight of the tributary area and H is the total height.

Mode j (by neglecting shear deformations):

$$\phi_j = A_j\left[\sin C_j x - \sinh C_j x + \frac{\sin C_j H + \sinh C_j H}{\cos C_j H + \cosh C_j H} \times (\cosh C_j x + \cos C_j x)\right]$$

(A.28)

$$C_j^4 = \frac{m\omega_j^2}{EI}$$

A_j = an arbitrary constant.

Periods:

$$T_1 = 1{\cdot}787\,T_0; \qquad T_2 = 0{\cdot}285\,T_0; \qquad T_3 = 0{\cdot}102\,T_0; \cdots;$$

$$T_j = \frac{1{\cdot}425}{(2j-1)}\,T_0; \cdots$$

where

$$T_0 = \sqrt{\left(\frac{WH^3}{gEI}\right)}$$

A.2.5 UNDAMPED FORCED VIBRATIONS

See Figure A.13. The set of equations of vibration can be 'decoupled' so that each mode will be described by a single equation.

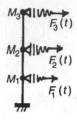

Figure A.13

(a) *Flexibility formulation*

We define, by referring to the mode of vibration j:

- Generalized mass:

$$M_j^* = \boldsymbol{\phi}_j^T \cdot \boldsymbol{M} \cdot \boldsymbol{\phi}_j = \sum_i M_i \phi_{i_j}^2 \qquad (A.29)$$

- Generalized exciting force:

$$F_j^*(t) = \boldsymbol{\phi}_j^T \cdot \boldsymbol{f}(t) = \phi_{i_j} F_1(t) + \phi_{2_j} F_2(t) + \cdots \left.\vphantom{\begin{array}{c}1\\1\end{array}}\right\}$$
$$+ \phi_{i_j} F_i(t) + \cdots + \phi_{n_j} F_n(t) \qquad (A.30)$$

where $\boldsymbol{f}(t)$ is the vector of exciting forces:

$$\boldsymbol{f}(t) = \begin{bmatrix} F_1(t) \\ F_2(t) \\ \vdots \\ F_n(t) \end{bmatrix} \qquad (A.31)$$

The set of equations of vibration becomes

$$M_j^* \ddot{Y}_j(t) + \omega_j^2 M_j^* Y_j(t) = F_j^*(t) \qquad (A.32)$$

(b) *Stiffness formulation*

We define the generalized rigidity

$$K_j^* = \boldsymbol{\phi}_j^T \cdot \boldsymbol{K} \cdot \boldsymbol{\phi}_j \qquad (A.33)$$

The generalized rigidity and the generalized mass are related by the equation

$$K_j^* = \omega_j^2 M_j^* \qquad (A.34)$$

The equations of motion become

$$M_j^* \ddot{Y}_j(t) + K_j^* Y_j(t) = F_j^*(t) \qquad (A.35)$$

A.2.6 DAMPED FORCED VIBRATIONS

Damping forces:

$$\left.\begin{aligned}
F_{D_1} &= b_{11}\dot{u}_1 + b_{12}\dot{u}_2 + \cdots + b_{1n}\dot{u}_n \\
F_{D_2} &= b_{21}\dot{u}_1 + b_{22}\dot{u}_2 + \cdots + b_{2n}\dot{u}_n \\
&\vdots \\
F_{D_n} &= b_{n1}\dot{u}_1 + b_{n2}\dot{u}_2 + \cdots + b_{nn}\dot{u}_n
\end{aligned}\right\} \quad (A.36)$$

Damping force vector:

$$\left.\boldsymbol{f}_D = \begin{bmatrix} F_{D_1} \\ F_{D_2} \\ \vdots \\ F_{D_n} \end{bmatrix}\right.$$

Velocity vector:

$$\left.\dot{\boldsymbol{u}} = \begin{bmatrix} \dot{u}_1 \\ \dot{u}_2 \\ \vdots \\ \dot{u}_n \end{bmatrix}\right\} \quad (A.37)$$

Damping matrix:

$$\left.\boldsymbol{B} = \begin{bmatrix} b_{11} & b_{12} & \cdots & b_{1n} \\ b_{21} & b_{22} & \cdots & b_{2n} \\ \vdots & & & \vdots \\ b_{n1} & b_{n2} & \cdots & b_{nn} \end{bmatrix}\right.$$

Equations of motion:

$$\boldsymbol{f}_D = \boldsymbol{B} \cdot \dot{\boldsymbol{u}} \quad (A.38)$$

Equation j:

$$M_j^* \ddot{Y}_j(t) + B_j^* \dot{Y}_j(t) + K_j^* Y_j(t) = F_j^*(t) \quad (A.39)$$

where $B_j^* = 2\omega_j \xi_j M_j^*$ is the generalized damping coefficient. The property of decoupling of modes of vibration also remains valid for forced vibrations.

A.2.7 ARBITRARY EXCITATION

Solution of the equation of motion j for an arbitrary exciting force:

$$Y_j(t) = \frac{1}{M_j^* \omega_j} \int_0^t F_j^*(\tau) e^{-\omega_j \xi_j (t-\tau)} \sin \omega_j (t-\tau) d\tau \qquad (A.40)$$

A.3 Seismic forces

A.3.1 SINGLE-DEGREE-OF-FREEDOM STRUCTURES

The SDOF system shown in Figure 6.18 is acted upon by a random ground motion; the displacement u_G, the velocity $\dot{u}_G$ and the acceleration $\ddot{u}_G$ are given, as records of the ground motion during earthquake.

$$u_{TOT} = u_G + u$$

The equation of motion

$$\ddot{u} + 2\omega \xi \dot{u} + \omega^2 u = -\ddot{u}_G \qquad (A.41)$$

has the solution

$$u = \frac{1}{\omega} \mathscr{I}(t, \omega) = \frac{1}{\omega} \int_0^t \ddot{u}_G(\tau) e^{-\omega \xi (t-\tau)} \sin \omega (t-\tau) d\tau$$

$$\dot{u} = \frac{du}{dt} = \frac{1}{\omega} \frac{d\mathscr{I}(t, \xi)}{dt} = \frac{1}{\omega} \int_0^t \ddot{u}_G(\tau) [e^{-\omega \xi (t-\tau)} \omega \cos \omega (t-\tau)$$

$$- \omega \xi \sin \omega (t-\tau) e^{-\omega \xi (t-\tau)}] d\tau \qquad (A.42)$$

The term $\omega \xi \sin \omega (t-\tau) e^{-\omega \xi (t-\tau)}$ is usually negligible. Hudson (1962) has shown that we may replace in the expression for velocity $\dot{u}$ the term $\cos \omega (t-\tau)$ by $\sin \omega (t-\tau)$. Consequently the expression for the velocity becomes

$$\dot{u} \cong \int_0^t \ddot{u}_G e^{-\omega \xi (t-\tau)} \sin \omega (t-\tau) d\tau \qquad (A.43)$$

and $\dot{u} \cong u/\omega$

According to Newmark and Rosenblueth (1971) this approximation affects significantly the maximum velocities only for very long periods and the maximum accelerations only for very short periods.

The response spectra result as:

● S_d, the maximum relative displacement recorded for a given earthquake in a specific direction:

$$S_d = u_{max} = \frac{1}{\omega} \mathscr{I}(t, \xi) \qquad (A.44)$$

• S_v, the maximum relative velocity:

$$S_v = \dot{u}_{max} = \mathscr{I}(t, \xi)_{max} = \omega S_d \tag{A.45}$$

• S_a, the maximum total acceleration:

$$S_a = \ddot{u}_{TOT_{max}} = \omega \mathscr{I}(t, \xi)_{max} = \omega^2 S_d = \omega S_v \tag{A.46}$$

The maximum forces:

$$F_{max} = M \ddot{u}_{TOT_{max}} = M S_a = M \omega S_v = M \omega^2 S_d \tag{A.47}$$

The velocity spectrum S_v and the acceleration spectrum S_a obtained by using the aforementioned approximation are sometimes called the pseudo-velocity spectrum and pseudo-acceleration spectrum.

A.3.2 MULTI-DEGREE-OF-FREEDOM STRUCTURES

The effective mass M_j^e and effective weight W_j^e are given by

$$M_j^e = \frac{\left(\sum_i M_i \phi_{i_j} \right)^2}{\sum_i \left(M_i \phi_{i_j}^2 \right)}$$

$$W_j^e = M_j^e g = \frac{\left(\sum_i W_i \phi_{i_j} \right)^2}{\sum_i \left(W_i \phi_{i_j}^2 \right)} \tag{A.48}$$

The sum of the effective masses (weights) is equal to the total mass M (weight W).

The maximum modal force $F_{i_{j_{max}}}$ at storey i, corresponding to the mode j, results in

$$F_{i_{j_{max}}} = \frac{M_i \phi_i \sum_i M_i \phi_{i_j}}{\sum_i M_i \phi_{i_j}^2} S_{a_j} = \frac{W_i \phi_i \sum_i W_i \phi_{i_j}}{g \sum_i W_i \phi_{i_j}^2} S_{a_j} \tag{A.49}$$

The maximum base shear (sum of forces $F_{i_{j_{max}}}$):

$$V_{j_{max}} = \sum F_{i_{j_{max}}} = \frac{\left(\sum_i M_i \phi_{i_j} \right)^2}{\sum_i \left(M_i \phi_{i_j}^2 \right)} S_{a_j} = M_j^e S_{a_j} = \frac{W_j^e S_{a_j}}{g} \tag{A.50}$$

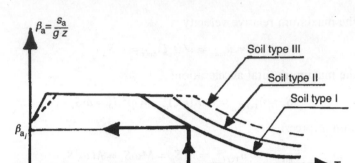

Figure A.14

The force $F_{i_{j_{max}}}$ acting at level i can be expressed in the form:

$$F_{i_{j_{max}}} = d_{i_j} V_{j_{max}} \qquad (A.51)$$

where the distribution coefficient d_{i_j} is

$$d_{i_j} = \frac{M_i \phi_{i_j}}{\sum_i M_i \phi_{i_j}} = \frac{W_i \phi_{i_j}}{\sum_i W_i \phi_{i_j}} \qquad (A.52)$$

Usually, the codes prescribe spectra β_a versus T (Figure A.14), where $\beta_a = S_a/a_{G_{max}} = S_a/gZ$; $a_{G_{max}}$ denotes either the peak ground acceleration or the effective ground acceleration (see section 6.1.2); usually several spectra are drawn, depending on the type of soil. In order to take into consideration the importance of the building and the structural behaviour (mainly the effect of ductility), additional factors are introduced $(I, 1/R)$. The base shear becomes

$$V_j = \frac{W_j^e S_{a_j} I}{g R} = \frac{W_j^e \beta_{a_j} Z I}{R} = c_j W_j^e \qquad (A.53)$$

The seismic coefficient c_j for the mode j results in

$$c_j = \frac{\beta_{a_j} Z I}{R} \qquad (A.54)$$

β_{a_j} is taken from the given code spectrum, for the period T_j and the considered type of soil.

As the maximum modal forces $F_{i_{1_{max}}}, F_{i_{2_{max}}}, \ldots, F_{i_{n_{max}}}$ do not occur simultaneously, their effects r are superimposed 'statistically':

$$r = \sqrt{(r_1^2 + r_2^2 + \cdots + r_h^2)} \qquad (A.55)$$

The number of modes h to be taken into consideration depends on the required accuracy; usually

$$W_1^e + W_2^e + \cdots W_h^e \geqslant 0 \cdot 9\, W \qquad (A.56)$$

where W is the total weight of the structure and $W_1^e, W_2^e, \ldots, W_h^e$ denote the effective weights of the modes $1, 2, \ldots, h$.

The total seismic force (base shear) yielded by modal analysis (V_{mod}) is usually less than the total seismic force yielded by the static lateral force procedure (V_{st}). Modern codes (e.g. SEAOC-88) require in such cases to multiply the results obtained by modal analysis by a factor intended to reduce this gap: $(0.8 \ldots 1.0) \times (V_{st}/V_{mod})$ – the so called "scaling of results".

Bibliography

Anderson, J. (1989) Dynamic response of buildings, in *The Seismic Handbook* (ed. F. Naeim), Van Nostrand Reinhold, New York, Ch.3.

Blume, J., Newmark, N. and Corning, L. (1961) *Design of Multistorey RC Buildings*, Portland Cement Association, Skokie, IL.

Capra, A. and Davidovici, V. (1980) *Calcul dynamique des structures en zone sismique*, Eyrolles, Paris.

Clough, R. and Penzien, J. (1993) *Dynamics of Structures*, McGraw-Hill, New York.

Davidovici, V. (ed.) (1985), *Génie parasismique*, Presses de l'École Nationale des Ponts et Chaussées, Paris.

Dowrick, D. (1987) *Earthquake Resistant Design*, J. Wiley, Chichester.

Hudson, D. (1962) Some problems in the application of spectrum techniques to strong-motion earthquake analysis. *Bulletin of the Seismological Society of America*, **52** (2), 417–430.

Mazilu, P. (1968) Curs de dinamica structurilor. Fac. Constructii (in Romanian).

Newmark, N. and Rosenblueth, E. (1971) *Fundamentals of Earthquake Engineering*, Prentice-Hall, Englewood Cliffs.

Scarlat, A. (1986) *Introduction to Dynamics of Structures*, Institute for Industrial and Building Research, Tel Aviv (in Hebrew).

SEAOC (1988) Seismic Committee – Structural Engineering Association of California.

Appendix B
Techniques for finite element computations

B.1 Approximate analysis of structural walls by finite elements

The finite element technique allows the most accurate analysis of coupled structural walls, provided a sufficiently dense mesh is used. In order to simplify the input work the mesh has to be regular – preferably with rectangular elements. It is recommended that the elements be nearly quadratic; ratios of length/width should be in the range 1/2 to 2/1.

The accuracy is usually proportional to the number of elements, but so are the volume of the input and the computer time. In order to assess the order of magnitude of the number of elements to be used in the analysis of coupled structural walls with a reasonable accuracy, we refer to the structure shown in Figure B.1a.

We choose a network that gives results that are not significantly improved by a further increase in the number of elements; we define the corresponding analysis as 'accurate'. In the specific example displayed in Figure B.1a a total of 760 elements has been used; the lintels ($2 \cdot 00 \times 0 \cdot 80$) have been divided into eight elements, each: $(4 \times 0 \cdot 60) \times (2 \times 0 \cdot 40)$ (Figure B.1b.)

A simplification of the analysis may be achieved in the following ways.

- Use of a coarser mesh, as shown in Figure B.1c, with a total of only 210 elements. The maximum differences with respect to the 'accurate' results are 10–20%.
- Use of a coarse mesh for the entire structure except for a given zone L, where we need more accurate results, and where we have used a denser mesh (Figure B.1d); within this zone the results are very close to the 'accurate' results.
- Replacing the lintels by horizontal beams as defined in Figure B.1e (the replacing beams have the same geometrical characteristics as the given lintels). Obviously, greatest accuracy is obtained when the lintels are shallow; then the results are nearly identical to the 'accurate' ones.

B.2 Use of symmetry properties

In the case of symmetric structures subjected to symmetric or anti-symmetric loads we can perform the computations on a segment of the real structure only, contained between the planes of symmetry (the 'structural segment' S^*). It obviously leads to a significant decrease in the computation volume.

We must ensure that the elastic line of the considered segment S^* along the plane of symmetry preserves the shape of the elastic line of the original structure.

To this end we have to use the following means.

- Proper **symmetry supports** must be introduced in the nodes lying on the planes of symmetry. They are meant to ensure that their displacements preserve the type of displacements imposed by the symmetry conditions (the symmetry supports must restrain the directions of known zero displacements).
- The geometrical characteristics of the elements lying on the symmetry plane must be adjusted in agreement with the conditions of symmetry.

By referring to the symmetric structure shown in Figure B.2a, when symmetrical loads are considered, the symmetry support displayed in Figure B.2b must be introduced (the conditions of symmetry impose zero displacements (rotations) in directions $X1$ and $X6$). When anti-symmetric loads are considered, the symmetry support displayed in Figure B.2c fits the conditions of symmetry (zero displacement in direction $X2$).

By referring to the symmetric structure shown in Figure B.3a,where the central column lies on the plane of symmetry, we have to adjust its area A and moment of inertia I in order to ensure that the elastic line of the considered segment is in agreement with the elastic line of the original structure; this means that the corresponding column of the considered fragment has an area $A/2$ and a moment of inertia $I/2$ (Figure B.3b,c).

The proper symmetry supports and the needed adjustments of the geometrical characteristics for any symmetric structure are detailed in Table B.1.

An element lying on the plane of symmetry will have the following **adjusted geometrical characteristics**: area $A^* = A/2$; moment of inertia (bending) $I^* = I/2$; moment of inertia (torsion) $I_t^* = I_t/2$. We have to bear in mind that the symmetry supports have to be added to the existing supports of the original structure.

Consequently, the fixed ends of columns and the pinned supports of finite elements of the structural walls remain unchanged when defining the structural segment S^*.

In the case of springs lying on the symmetry plane and symmetrical loads we have to adjust the corresponding spring constant k: $k^* = k/2$.

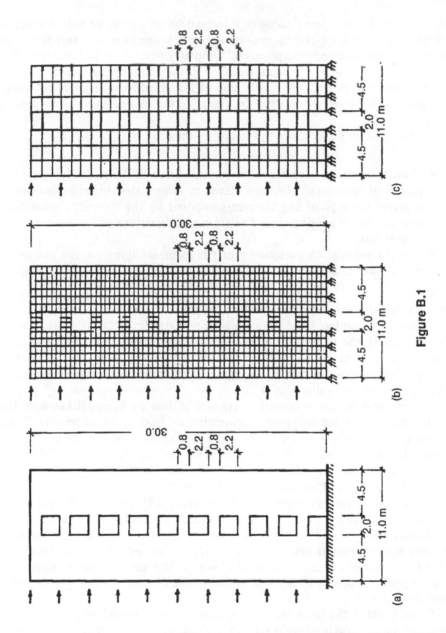

Figure B.1

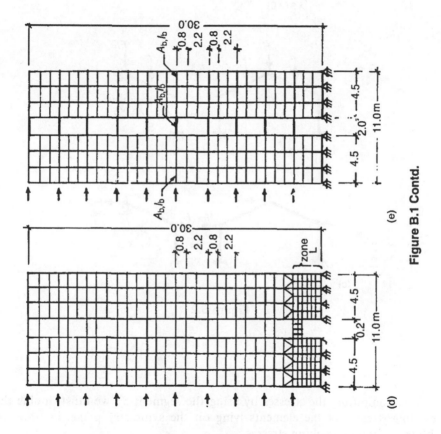

Figure B.1 Contd.

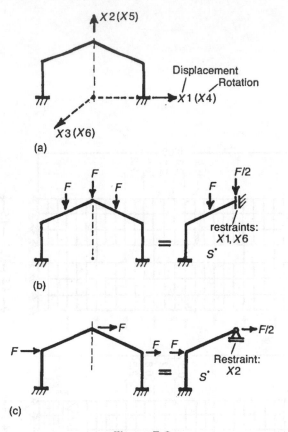

Figure B.2

After computing the stresses by using the segment S^* we must double the resulting stresses of the elements lying on the symmetry plane, in order to obtain the actual resulting stresses.

☐ **Numerical example B**

The space structure shown in Figure B.4a is symmetrical with respect to two planes: $X1–X3$ and $X2–X3$. The given loads are symmetrical with respect to the plane $X1–X3$ and anti-symmetrical with respect to the plane $X2–X3$.

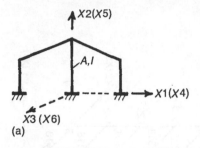

(a)

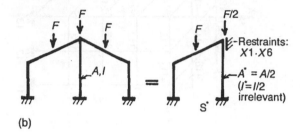

(b)

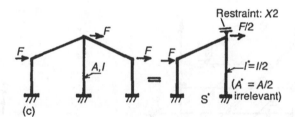

(c)

Figure B.3

The computations can be performed on the symmetrical segment S^* shown in Figure B.4b. According to the data displayed in Table B.1 the restraints are as follows:

- $X1$, $X2$, $X3$, $X4$, $X5$, $X6$: nodes 11–23 (fixed ends in the original structure: no adjustment is needed for the considered segment S^*);
- $X2$, $X4$, $X6$: nodes 11–411 by 100, 12–412 by 100 (symmetry with respect to $X1$, $X3$);
- $X2$, $X3$, $X4$: nodes 18–418 by 100, 23–423 by 100 (anti-symmetry with respect to $X2$, $X3$);
- $X2$, $X3$, $X4$, $X6$: nodes 13–413 by 100 (symmetry with respect to $X1$, $X3$ and anti-symmetry with respect to $X2$, $X3$).   □

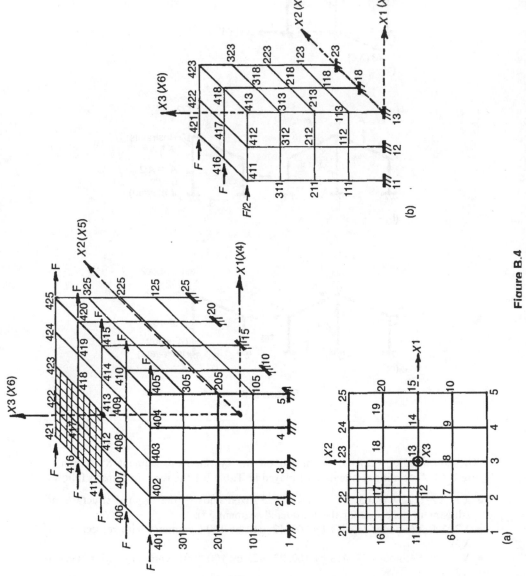

Figure B.4

Table B.1 Symmetry supports

Plane of symmetry			Symmetrical loads	Antisymmetrical loads
Plane Structure (X1, X2)		$X1, X3$	$X2, X6$	$X1$
		$X2, X3$	$X1, X6$	$X2$
Grid Structure (X1, X2)		$X1, X3$	$X4$	$X3, X5$
		$X2, X3$	$X5$	$X3, X4$
Space Structure		$X1, X2$	$X3, X4, X5$	$X1, X2, X6$
		$X1, X3$	$X2, X4, X6$	$X1, X3, X5$
		$X2, X3$	$X1, X5, X6$	$X2, X3, X4$

Appendix C
Methods of quantifying the structural rigidity

Two main factors must be taken into account when we look for the order of magnitude of the effective rigidities of structures subject to lateral loads: the effect of soil deformability and the effect of cracking of reinforced concrete structures. These factors strongly affect the magnitude of seismic forces and their distribution between structural walls and frames, as well as the thermal stresses.

C.1 Quantifying the effect of soil deformability

The effect of soil deformability on the deflections/rigidities of structural walls is very important, and overlooking it may lead to significant errors.

C.1.1 SPREAD AND MAT FOUNDATIONS

(a) From a theoretical point of view, the most accurate method for determining the effect of soil deformability is the so-called **interaction analysis**, in which the structure, the foundations and the surrounding soil are dealt with as a single system (Figure C.1). The extent of the soil (the sizes L_s, B_s, H_s) is chosen so that the stresses on its periphery are negligible; consequently, the supports along this periphery have no practical effect on the analysis. Obviously, such an analysis has to be performed by tri-dimensional ('solid') finite elements only. We must choose the moduli of elasticity E_s (Young modulus) for normal stresses and G_s for shear stresses (or the Poisson's ratio) of the soil.
   This method has several drawbacks:

- The computation involves a very high number of unknowns and, as such, needs special computer capability.
- The validity of the usual assumption of elastic/uncracked soil elements is rather questionable; we may partly correct the effect of this assumption by a non-linear analysis that also includes gaps, but additional computational complications are involved.
- The soil is very non-homogeneous, and we cannot overcome this intricacy by computational means.

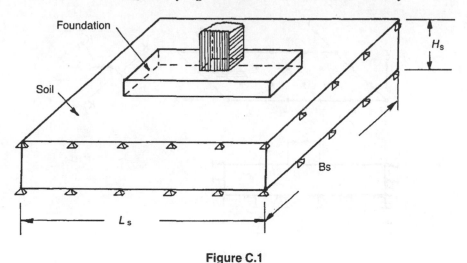

Figure C.1

- The recommended moduli of elasticity of the soil are uncertain, and this strongly affects the results.
- No experimental confirmation of interaction analyses is available, especially of the frail assumptions made in a non-linear analysis; computations have shown that the corresponding rigidity of the soil may decrease by 50% with respect to the rigidity resulting from an elastic analysis.

Consequently, when considering an elastic soil, it is advisable to consider two values of the modulus of elasticity of the soil – a high value and a low value – and to perform two corresponding computations; the design will be based on the highest stresses.

(b) A simpler method to quantify the effect of soil deformability, albeit less accurate from a theoretical point of view, is based on substituting a set of discrete elastic springs (Figure C.2) for the continuous, deformable soil.
    The spring constants are proportional to the subgrade modulus of the soil, k_s (kN m^{-3}) and the tributary area A_i (m^2):

$$k_i = k_s A_i \qquad\qquad (C.1)$$

This procedure is based on Winkler's assumption, and as such it neglects the interaction of adjacent springs; the errors increase in the case of soft soils.

(c) If we assume that the foundation base remains plane after the soil's elastic deformation, we also implicitly admit a linear variation of stresses.

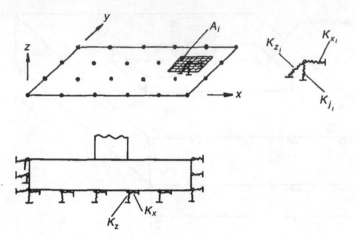

Figure C.2

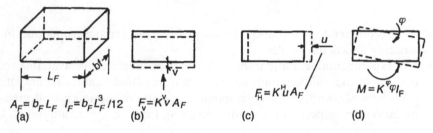

Figure C.3

Consequently, we can replace the set of elastic springs by three 'global' springs in the centre of the foundation (Figure C.3) with the constants

K^v (kN m^{-1}) for vertical displacements;
K^H (kN m^{-1}) for horizontal displacements;
K^φ (kN m rad^{-1}) for rotations;

defined as follows:

$$\left. \begin{aligned} v &= F^v/K^v; & K^v &= k_s^v A_f \\ u &= F^H/K^H; & K^H &= k_s^H A_f \\ \varphi &= M/K^\varphi; & K^\varphi &= k_s^\varphi I_f \end{aligned} \right\} \tag{C.2}$$

Usually, we assume $k_s^V = k_s^H = k_s^\varphi = k_s$, although several tests have shown different values (Barkan, 1962; SNiP 2.02.03, 1985). The magnitude of the subgrade moduli vary with the sizes and the form of the foundation (Bowles, 1993).

It is worth noting that the method based on the assumption of linear elastic stresses, which is the least accurate from a theoretical point of view, is the only one based on tests and therefore, in fact, the most reliable. It is also consistent with the usually accepted assumption that the design computations are performed on the basis of linear elastic stresses. In this context we would like to quote from the *Seismic Design Handbook* (Naeim 1989)

... arriving to a mathematical model to describe the inelastic behaviour of structures during earthquake is a difficult task... Procedures for utilizing inelastic spectra in the analysis and design of multi-degree-of-freedom systems have not yet been developed to the extent that they can be implemented in design (Mohraz and Elghadamsi, p. 77). Great care must be exercised in soil–structure interaction analyses... The current state of art falls far short of modelling reality (Lew and Nissen, p. 375).

Paulay and Priestley (1992) recommend considering the columns of moment-resisting frames modelled for seismic design as rotational springs (except for columns supported on raft foundations or individual pads supported by short stiff piles or by foundation walls in basements); when gravity loads are considered, fixed ends are recommended.

We consider it advisable to use sets of discrete springs for large foundations and 'global' central springs for small foundations. In the numerical examples of the present book two types of soil have been considered: soft soils (subgrade modulus $k_s = 20\,000$ kN m^{-3}) and hard soils (subgrade modulus $k_s = 100\,000$ kN m^{-3}). Several relationships have been proposed between subgrade moduli k_s and moduli of elasticity E_s. Calibration computations have shown that, by considering elastic soil elements, subgrade moduli of $20\,000$–$30\,000$ kN m^{-3} correspond roughly to moduli of elasticity of $40\,000$–$60\,000$ kN m^{-2}, while subgrade moduli of $80\,000$–$100\,000$ kN m^{-3} correspond roughly to moduli of elasticity of about $200\,000$ kN m^{-2}.

C.1.2 FOUNDATIONS ON PILES

Contrary to the prevailing opinion, foundations on piles do not ensure a high degree of fixity. Significant horizontal as well as vertical deformations develop, and we have to take them into account.

(a) Horizontal deflections

In order to compare the horizontal deformability of piles and spread footings, computations have been performed for piles of 10 m length and footings with a similar vertical bearing capacity, subjected to identical horizontal forces, and the corresponding horizontal deflections were computed. We have considered, in agreement with the Russian code SNiP 2.02.03 (1985), an equivalent soil strip of 1·10 m for piles with 0·40 m diameter and 2·50 m for piles with 1·50 m diameter. Denoting by $K_p(K_f)$ the horizontal rigidities of the piles (spread footings) the following results have been obtained:

Diameter of piles (m)	K_p/K_f
0·40	1/5 to 1/2
1·50	1/2·5 to 1/1·5

This shows that, from the point of view of horizontal deflections, the piles are much more deformable than the equivalent spread footings.

(b) Vertical settlements and rotations

In order to assess the order of magnitude of the vertical rigidity of piles, we shall assume that, at service load, a settlement of 0·3–0·5% of the pile diameter D_p is to be expected for reinforced concrete piles (Meyerhof, 1976; Poulos, 1980). Assuming that the piles are usually loaded at 80% of the maximum service load only, we obtain

$$V = \alpha D_p \tag{C.3}$$

where $\alpha = 0·25–0·40\%$.

Let us consider an **equivalent spread footing** where the normal stress is σ. Usually, the allowable stress is $\sigma_a = 100–400$ kN m^{-2}, so that we may admit a usual service stress $\sigma = 80–320$ kN m^{-2}. The corresponding settlement results in

$$v = \frac{\sigma}{k_s} \tag{C.4}$$

Equating (C.3) and (C.4) yields the **equivalent subgrade modulus**:

$$k_s^{eq} = \frac{\sigma}{\alpha D} \tag{C.5}$$

In the case of small pile diameters ($D_p = 0·40$ m), we obtain $k_s^{eq} = 50\,000–320\,000$ kN m^{-1}; somewhat greater than the usual subgrade moduli obtained for spread footings.

In the case of large pile diameters ($D_p = 1·50$ m), we obtain $k_s = 13\,000–85\,000$ kN m^{-3}: i.e. the same order of magnitude as for spread footings.

Hence the order of magnitude of the piles' rigidities is rather close to the values found for spread footings, and consequently we can accept the conclusions obtained for spread footings as qualitatively valid for pile foundations, too.

C.2   Quantifying the effect of cracking of RC elements

The effect of cracking in RC structures depends on the type of the structural element, the reinforcement ratio and the stresses level. Chopra and Newmark (1980) propose to evaluate the rigidity of RC elements by considering

an average of the moments of inertia... between cracked and completely un-cracked sections or as between net and gross sections..., unless they clearly are stressed at such low levels or that shrinkage is so limited that cracking is not likely.

The global effect of cracking on the rigidity can be described by the ratio I_e/I_g, where I_g is the moment of inertia of the uncracked element, and I_e is an equivalent moment of inertia, determined so that it will yield deflections close to the deflections of the cracked element. This ratio can be determined either by tests or by computations using finite element techniques, which take into account the effect of cracking. In the following we shall use data provided by Paulay and Priestley (1992).

C.2.1 MOMENT-RESISTING FRAMES

The ratio I_e/I_g varies between the values

$$\text{Rectangular beams: } 0.3\text{--}0.5$$

$$\text{T and L beams: } 0.25\text{--}0.45$$

Columns subject to high compressive strength:

$$\left(\sigma = \frac{N}{A} > 0.5 f_{ck} \right): 0.7\text{--}0.9$$

Columns subject to low compressive strength:

$$(\sigma = 0.2 f_{ck}): 0.5\text{--}0.7$$

where A is the cross-sectional area of the uncracked column, and f_{ck} denotes the characteristic (cube) strength of the concrete. Computations based on these values show that for the whole structure, a rough preliminary value of the ratio $I_e/I_g \cong 0.5$ can be assumed.

C.2.2 STRUCTURAL RC WALLS WITHOUT OPENINGS

The proposed ratio is

$$\frac{I_e}{I_g} = \frac{100}{f_y} + \frac{\sigma}{f_{ck}} \quad \text{(MPa)} \tag{C.6}$$

Computations based on this formula show that a rough preliminary value of the ratio $I_e/I_g \cong 0.5$ can be assumed.

C.2.3 STRUCTURAL RC COUPLED WALLS

The proposed ratio I_e/I_g depends on the type of shear reinforcement of the coupling beams.

When regular reinforcement is used (stirrups):

$$\frac{I_e}{I_g} = \frac{0.2}{1 + 3(h_b/l_n)^2} \tag{C.7}$$

When diagonal reinforcement is used:

$$\frac{I_e}{I_g} = \frac{0.4}{1 + 3(h_b/l_n)^2} \tag{C.8}$$

where l_m is the clear span and h_b is the height of the coupling beams.

Computations based on these formulae show that rough preliminary values of the ratio $I_e/I_g \cong 0.2$ (regular reinforcement) and $I_e/I_g \cong 0.4$ (diagonal reinforcement) can be assumed. Interesting information dealing with structural behaviour during earthquakes can be obtained by comparing the natural periods of buildings before and after major earthquakes. Ogawa and Abe (1980) compared the periods of more than 200 buildings in Sendai, Japan, before and after an earthquake of magnitude 6.8 and maximum ground accelerations of 0.25–0.40 g; the damage due to the earthquake led to increases in periods of about 31%. Similar comparisons were made in Leninakan, Armenia, before and after the 1988 earthquake. Results cited by Eisenberg (1994) displayed period increases of 50–70% for three buildings and an exceptional increase of 230% for a building relying on moment-resisting frames.

By assuming an increase of about 40% of the natural period and accepting the evaluation of seismic forces prescribed by the SEAOC-88 code we obtain a decrease of seismic forces of about 20% during earthquakes.

C.3   Final remarks

We have to distinguish between two types of deformation analyses of structures.

(a) We aim at determining the effective behaviour of a structure or of a specific structural element loaded by seismic forces corresponding to a given accelerogram (i.e. a **time history analysis**, where the structural response results from a numerical integration of the equations of motion).

As was shown in section 6.1.4, comparative analyses have been performed for a 20-storey RC moment-resisting frame and for a 20-storey RC structural wall. Both computations reached the same conclusion: the maximum deflections computed by an elastic, linear analysis and by an elasto-plastic, non-linear analysis are similar. The result can be explained by the decrease in rigidity due to cracking, a decrease that has two opposite effects: on the one hand it increases the displacements; on the other hand, it leads to larger periods and subsequently to a decrease in seismic forces accompanied by a corresponding decrease in displacements. The aforementioned analyses have shown that these opposite effects are nearly equal, and therefore the deflections yielded by a non-linear, elasto-plastic analysis and by an elastic, linear analysis are nearly equal.

This important conclusion allows us to use in design the simple elastic, linear analysis.

(b) We aim at evaluating the **real rigidities** of various components of an RC structure in order to obtain a correct force distribution among structural elements. In this case, according to the definition of the rigidity, we have to choose a given pattern of horizontal forces with an arbitrary intensity (usually we choose either a force $F = 1$ concentrated at the top of the building, or a set of triangularly distributed loads with a top load $F_{max} = 1$) and to compute the maximum deflection u_{max} by an elastic analysis; then we define the rigidity as $K = 1/u_{max}$. When we compute the rigidities of the structural elements in order to obtain the maximum deflection u_{max}, we can use the aforementioned recommendations for decreasing them due to cracking.

We note that the effects of soil deformability and cracking are interdependent: the soil deformability leads to a decrease in rigidity and seismic forces and therefore the effect of cracking will decrease, and similarly the cracking leads to a decrease in rigidity and a corresponding decrease of the effect of soil deformability. It is not allowed to take into consideration a direct cumulative effect of soil deformability and cracking.

Bibliography

Barkan, D. (1962) *Dynamics of Bases and Foundations*, McGraw-Hill, New York (translated from Russian).

Bowles, G. (1993) *Foundation Analysis and Design*, McGraw-Hill, New York.

Burland, J., Butler, F. and Duncan, P. (1966) The behaviour and design of large-diameter bored piles in soft clay, in *Proceedings of the Symposium on Large Bored Piles*, Institution of Civil Engineers, London, pp. 51–72.

Chopra, A. and Newmark, N. (1980) Analysis, in *Design of Earthquake Resistant Structures* (ed. E. Rosenblueth), Pentech Press, London, Ch. 2.

Darragh, R. and Bell, R. (1969) *Load Tests on Long Bearing Piles*, ASTM, Publ. 444, American Society for Testing and Materials, Philadelphia.

Eisenberg, J. (1994) Lessons of recent large earthquakes and development of concepts and norms for structural design, in *Proceedings of the 17th European Regional Earthquake Engineering Seminar*, Technion, Haifa, 1993, Balkema, Rotterdam, pp. 29–38.

Meyerhof, G. (1976) Bearing capacity and settlements of pile foundations. *Inst. Geotechnical Engineering Division ASCE*, **102**(GT3), 195–228.

Naeim, F. (ed.) (1989) *Seismic Design Handbook*, Van Nostrand, New York.

Ogawa, J. and Abe, Y. (1980) The stiffness degradation caused by a severe earthquake, in *Proceedings of International Conference on Earthquake Engineering for Protection from Natural Disasters*, Bangkok, pp. 39–50.

Paulay, T. and Priestley, M. (1992) *Seismic Design of Reinforced Concrete and Masonry Buildings*, J. Wiley & Sons, New York.

Poulos, H. and Davies, A. (1980) *Pile Foundation Analysis and Design*, J. Wiley, New York.

Scarlat, A. (1985) Design for temperature changes, prejudices and reality, in *Third Symposium of the Israeli Association of Civil Engineers*, Jerusalem, pp. 33. 1–20 (in Hebrew).

Scarlat, A. (1993) Effect of soil deformability on rigidity-related aspects of multistory buildings analysis. *ACI Structural Journal*, **90**(March/April), 156–162.

SEAOC (1988) *Recommended lateral forces requirements and tentative commentary*, Seismology Committee, Structural Association of California, San Francisco.

SNiP (1985) 2.02.03-85 -Svaennie fundamenti, Moscow (in Russian), 1985.

Appendix D
Glossary of earthquake engineering terms

accelerogram Diagram of acceleration versus time recorded during an earthquake (real or simulated). By integration we can obtain the corresponding velocity and displacement diagrams.

amplification characteristics of surface layers Amplification of the amplitude of a seismic wave when it propagates from the base rock towards the surface layers.

aspect ratio of a wall Height/length ratio.

attenuation Characteristic decrease in amplitude of the seismic waves with distance from source, due to geometric spreading of propagating wave, energy absorption and scattering of waves.

base The level at which earthquake motions are considered to be imparted to the structure.

base shear The shear force at the base of a structure (equal to the total lateral force).

bearing wall An interior or exterior wall providing support for vertical loads.

bedrock A rock that is not underlain by unconsolidated materials.

braced frame A vertical truss system provided to resist lateral forces (concentric or excentric).

building Low rise: one to three storeys
　　　　　Medium rise: four to seven storeys
　　　　　High rise: eight storeys and taller
　　　　　- according to classification proposed by the Federal Emergency Management Agency (FEMA) 154/1988.

building frame system A structural system with an essentially complete space frame providing support for vertical loads. Seismic resistance is provided by structural walls or braced frames.

centre of mass (of a given slab) The centre of gravity of masses (structural and non-structural) on the slab.

centre of rigidity (of a given slab) A point having the following property: when a horizontal force passes through it, the slab displays only translational motion (we assume that the slabs are rigid in both horizontal and vertical planes and that the vertical structural elements are inextensible). For the general case (deformable structural slabs and vertical elements) several definitions have been proposed (see Chapter 4).

circular frequency The frequency multiplied by 2π (rad s^{-1}).

collector An element provided to transfer lateral forces from a portion of a structure to the vertical elements of the lateral resisting system.

concentric braced frame A braced frame in which the members are subjected primarily to axial forces.

confined region The portion of an RC component in which the concrete is confined by closely spaced special lateral reinforcement restraining the concrete in directions perpendicular to the applied stresses.

cycle The motion completed during a period.

damping Internal viscous damping: damping associated with material viscosity, proportional to velocity. Body friction damping (Coulomb damping): damping taking place at connections and supports. Hysteresis damping: damping taking place when a structure is subjected to load reversal in the inelastic range.

degree of freedom (in structural dynamics, by considering a system with lumped masses) The number of independent parameters needed to define the position of the lumped masses. Frequently, only the flexural deformations are taken into account (the axial and shear deformations being neglected). Two main types of structure are usually referred to: single-degree-of-freedom (SDOF) and multi-degree-of-freedom (MDOF).

diaphragm A horizontal system designed to transmit lateral forces to vertical resisting elements of the system.

dual system A structural system with an essentially complete space system to provide support for gravity loads. The resistance to lateral loads is provided by structural walls and moment-resisting frames. The two systems are designed to resist the total lateral load in proportion to their relative rigidities; the moment-resisting frames are designed to resist a prescribed percentage of the total lateral force.

ductility Property of material, element or structure subjected to cyclic loads, to sustain large inelastic deformations before failure. The ductility is quantified by the **ductility ratio**, usually the ratio of a displacement (or curvature) at failure, to the same displacement (or curvature) at yield point. Ductility is usually included in **reduction factors**, prescribed by various codes.

duration of preliminary tremors (T_{ps}) The time interval between the arrival at the observation station of P and S waves. Usually, $T_{ps} \cong 0.12\, D$ s, where D is the distance from the hypocentre to the observation point, in km.

dynamic analysis (elastic or inelastic) Analysis based either on modal analysis or on direct integration of the equations of motion by using a step-by-step technique; in this latter case an appropriate seismic accelerogram is considered as input data.

eccentrically braced frame A braced frame with excentric connections, where significant bending moments develop, besides axial forces.

eccentricity The distance between the centre of mass and the centre of rigidity projected on a direction normal to the considered horizontal force.

epicentre (epifocus) The projection of the hypocentre onto the surface of the earth.

equivalent static lateral force procedure A method that replaces seismic lateral forces by static lateral forces.

exceedance probability The statistical probability that a specified level of ground motion will be exceeded during a specified period of time.

frequency The number of complete cycles in a unit of time ($Hz = cycles\,s^{-1}$); equal to 1/period.

hypocentre (focus) The point where the earthquake originates.

intensity scale The scale of ground-motion intensity as determined by human feelings and by the effect of ground motion on structures. The most used: Modified Mercalli (MM), 12 grades; Medvedev–Sponheuer–Karnik (MSK), 12 grades; Japanese Meterological Agency (JMA), 8 grades.

lateral force-resisting system A part of the structural system, assigned to resist lateral forces.

liquefaction Phenomenon whereby a saturated sandy layer loses its shear strength, owing to earthquake motion, and behaves like liquid mud.

limit state Serviceability limit state is reached when the building becomes unfit for its intended use through deformation, vibratory response, degradation or other physical aspects. Ultimate limit state is reached when the building fails, becomes unstable or loses equilibrium.

magnitude (M) A measure of the amount of energy released by an earthquake. According to a definition proposed by Ch. Richter (1934): $M = \log A$, where A (μm) is the maximum amplitude registered by a specific seismograph, at a point 100 km from the epicentre.

major damage Repairs would cost approximately 60% of the building's value (land or site improvement not included) – according to proposal of Federal Emergency Management Agency (FEMA) no.154/1988.

modal analysis Dynamic analysis of an n degree-of-freedom structure based on the decoupling of n single-degree-of-freedom deformed shapes (**modes of vibration**) and their superposition in order to reconstitute the deformed shape of the vibrating structure.

moment-resisting frame A frame in which the members are subjected primarly to bending moments.

moment ratio (of a structural wall) The ratio base moment of the wall / total overturning moment of the structure.

non-bearing wall An interior or exterior wall that does not provide support for vertical loads other than its own weight.

overturning moment (at a given storey) The moment of the resultant lateral forces above the considered storey with respect to an axis in the considered slab's plan, normal to the considered seismic forces. The corresponding axial forces in columns and shear walls must be added to the axial forces due to vertical forces.

$P\,\Delta$ effect Development of significant additional stresses due to large horizontal deflections of columns; they may be obtained by second-order analysis (the equilibrium is formulated by considering the deformed shape of the structure).

peak ground acceleration (PGA) The horizontal maximum acceleration during earthquakes on the free surface of a rock or a similar stiff soil, with a prescribed probability of being exceeded within a certain time span. Recently, the concept of **effective ground acceleration** (EGA) has been introduced, where a mean of low-frequency spikes of acceleration (having a more significant influence on the response and behaviour of structures) is considered. The zoning map of SEAOC-88 code is based on EGA.

period The time elapsed while the motion repeats itself (s). In modal analysis each mode has its own period; the first one (the longest) is the **fundamental period**.

radius of rigidity of a storey Square root of the ratio torsional rigidity/translational rigidity.

reduction factor A measure of the ability of a structural system to sustain cyclic inelastic deformations without collapse. It depends, besides the ductility ratio, on the existence of alternative **lines of defence**, such as lateral force system redundancy, non-structural elements, changed damping and period modification with deformation.

regular structure A structure not very slender and with no significant plan or vertical discontinuities of its resisting system.

reliability of a structure (in seismic design) The probability that the structure will survive all the actions exerted upon it by seismic forces during a given time interval.

response spectra Curves plotting the maximum values of relative displacements (S_d), relative velocities (S_v) or absolute accelerations (S_a) of a single-degree-of-freedom system versus its period, for a given damping ratio, according to recordings of earthquakes. In fact, slight simplifying modifications are allowed in the analytical expression for the maximum relative velocity and absolute acceleration, and hence we sometimes refer to spectrum of pseudo-velocity and spectrum of pseudo-acceleration (see Appendix A).

response spectrum analysis An elastic dynamic analysis of a structure, utilizing the peak dynamic response of all modes having a significant contribution to the total structural response.

rotational vibration Vibration involving rotations with respect to a horizontal axis (also: rocking vibration).

seismic coefficient The ratio total lateral seismic force/total vertical load.

seismic waves (main) (a) Longitudinal waves (**compression waves** P); (b) transverse waves (**shear waves** S); (c) Love waves (R) and (d) Rayleigh waves (**surface waves**).

seismic hazard The physical effects of an earthquake.

seismic resistant system The part of a structural system that is considered to provide the required resistance to seismic forces in design analysis.

seismic zone factor (Z) A dimensionless number mapped or defined in building codes that is roughly proportional to the ratio of a design earthquake acceleration to the acceleration of gravity.

shear ratio (of a structural wall) The ratio base shear taken by the wall/total base shear of the structure.

soft storey A storey in which the vertical resisting elements possess a lateral stiffness significantly lower than the corresponding stiffness of the storey above.

soil–structure interaction Analysis that considers the given structure (foundations included) and the surrounding soil as a 'complex structure' with different elastic–mechanical properties, subject to static or dynamic loads.

space frame A structural system composed of interconnected members other than bearing walls, that is capable of supporting vertical loads and that will also provide resistance to horizontal seismic forces.

storey drift The horizontal displacement of one level relative to the level below.

storey drift ratio The storey drift divided by the storey height.

storey torsional moment Moment of the storey shear multiplied by the eccentricity.

structural wall Wall proportioned to resist combinations of shears, moments and axial forces induced by lateral forces.

torsional rigidity of a storey The total moment of torsion due to torsional forces developing in the vertical resistant elements, as a result of a relative unit rotation of the slabs above and below the storey.

torsional vibration Vibration involving rotations with respect to a vertical axis.

translational rigidity of a storey Sum of shear forces developing in the vertical resisting elements in a specific direction when a relative unit displacement is impressed in the same direction.

weak storey A storey in which the total strength of the vertical resisting elements is significantly less than the corresponding stress of the storey above.

Appendix E
Design codes and standards

For more information about National Codes and Standards, the reader is referred to *International Handbook of Earthquake Engineering, Codes, Programs and Examples*, Edited by M. Paz, Chapman & Hall, 1995, 578 pp.

China

Chinese Academy of Building Research (1977) *Criteria for Evaluation of Industrial and Civil Buildings*, Beijing, TZ 23–77. English translation: Department of Civil Engineering, University of Washington, September 1984.

Canada

Canadian Prestressed Concrete Institute (CPCI) (1982) *Precast and Prestressed Concrete Design. Metric Design Manual*. Ottawa, Canada.

France

Association Française du Génie Parasismique (AFPS) (1990) *Récommendations AFPS-90 pour la rédaction de règles relatives aux ouvrages et installations a réaliser dans les régions sujettes aux séismes*, Presses ENPC, Paris.

Groupe de Coordination des Textes Techniques (DTU) (1982) *Règles Parasismiques 1969, Annexes et addenda 1982. (Règles PS 69, modifiée 82)* Editions Eyrolles, Paris, France.

Greece

Ministry for Environment and Public Works (1992) *Greek Code for Earthquake Resistant Structures*, Athens.

International

Comité Euro-International du Béton (CEB) (1985) *Model Code for Seismic Design of Concrete Structures*, Lausanne, April, Bulletin d'Information No. 165.

Commission of the European Communities (CEC) Technical Committee 250, SC8 (1994) *Eurocode 8. Earthquake Resistant Design of Structures: Part `1, General Rules and Rules for Buildings*, ENV 1998-1-1, CEN, Berlin.

Japan

Earthquake Resistant Regulations for Building Structures, Part 2, 1987. English translation in *Earthquake Resistant Regulations: A world list*, compiled by the International Association for Earthquake Engineering, July 1992.

New Zealand

Ministry of Works and Development (1981) *Pile Foundation Design Notes*, Civil Engineering Division, Wellington, New Zealand, CDP 812/8.

Standards Association of New Zealand (1976 and 1992) *Code of Practice for General Structural Design and Design Loadings for Buildings*, New Zealand Standards NZS 4203, Wellington, New Zealand.

Peru

National Institute of Research and Housing Regulations (ININVI) (1985) *Code for Aseismic Design, Norma E-020*, Cargas, Normas Técnicas de Edificación, Lima, Peru.

Romania

Institutul de Cercetare (1978, 1981, 1992) *Normativ Pentru Proiectarea Antiseismica a Constructiilor de Locuinte, Socio-culturale, Agro-zootechnice si Industriale* (Code for Earthquake Resistant Design of Dwellings, Socio-Economical, Agro-Zootechnical, and Industrial Buildings) P100-78, P100-81, P100-92, Bucuresti (in Romanian).

Spain

Presidencia del Gobierno, Comisión Interministerial (1974) *Norma Sismorresistente P.G.S.-1* (1974) Decreto 3209, Madrid, Spain.

United Kingdom

British Standards Institution (BSI) (1985) *Structural Use of Concrete*, London.

USA

American Association of State Highways and Transportation Officials (AASHTO) (1983) *Guide Specifications for Seismic Design*, Washington, DC, USA.

American Concrete Institute (ACI) (1988) *Building Code Requirements for Masonry Structures and Commentary*, Detroit, Michigan, USA, ACI 530-88, 550.1-88, ASCE 5-88, 6-88.

Applied Technology Council (ATC) (1974) *Model Code*, Redwood City, California.

Applied Technology Council (ATC) (1978) *Tentative Provisions for the Development of Seismic Regulations for Buildings*, Redwood City, California, ATC-3-06. National Bureau of Standards, Special Publication 510.

Applied Technology Council (ATC) (1988) *Rapid Visual Screening of Buildings for Potential Seismic Hazards: A Handbook*, Redwood City, California, ATC-21.

Applied Technology Council (ATC) (1989) *A Handbook for Seismic Evaluation of Existing Buildings*, Redwood City, California, ATC-22.

International Conference of Building Officials (ICBO) (1988 and 1991) *Uniform Building Code (UBC)*, Whittier, CA, USA.

International Conference of Building Officials (1992) *Uniform Code for Building Construction (UBC) Appendix, Chapter 1, Seismic Provisions for Unreinforced Masonry Bearing Walls and Buildings* (Advance version), Whittier, CA, USA.

National Center for Earthquake Engineering Research (NCEER) (1992) *A Procedure for the Seismic Evaluation of Buildings in the Central and Eastern United States*, by C. Polazides and J. Malley, State University of New York at Buffalo, Technical Report 92-0008.

National Earthquake Hazard Reduction Program (NEHRP) (1988) *NEHRP Recommended Provisions for the Development of Seismic Regulations for New Buildings*, Federal Emergency Management Agency (FEMA), Building Seismic Safety Council, Washington, DC, FEMA-97. pp. 1, 64, 65.

National Earthquake Hazard Reduction Program (NEHRP) (1992) *A Handbook for Seismic Evaluation of Existing Buildings*, Building Seismic Safety Council, Washington, DC.

NAVFAC P-335 *Seismic Design for Buildings*, TM 5-809-10. 1982.

Structural Engineers Association of California (SEAOC) (1990) *Recommended Lateral Force Requirements and Commentary*, Fifth Edition, Seismology Committee, Sacramento, California, USA.

Index